Lab Manual

Introduction to Electronics

Fourth Edition

Earl D. Gates

THOMSON LEARNING™

Australia Canada Mexico Singapore Spain United Kingdom United States

Introduction to Electronics Lab Manual
by Earl D. Gates

Business Unit Director:
Alar Elken

Acquisitions Editor:
Gregory L. Clayton

Development Editor:
Michelle Ruelos Cannistraci

Executive Editor:
Sandy Clark

Editorial Assistant:
Jennifer A. Thompson

Marketing Channel Manager:
Mona Caron

Production Manager:
Larry Main

Project Editor:
Christopher Chien

Executive Marketing Manager:
Maura Theriault

Art/Design Coordinator:
David Arsenault

Printed in Canada
1 2 3 4 5 XXX 05 04 02 01 00

For more information contact Delmar, 3 Columbia Circle, PO Box 15015, Albany, NY 12212-5015.

Or find us on the World Wide Web at http://www.delmar.com

Library of Congress Cataloging-in-Publication Data
00-064370
ISBN 0-7668-1700-8

NOTICE TO THE READER

TABLE OF CONTENTS

TABLE OF CONTENTS

PREFACE

This lab manual was written to enhance the textbook Introduction to Electronics: A Practical Approach. It provides a foundation for first year electronic students with the required hands-on experiences using real-world components and test equipment. It also provides an opportunity for students to develop a working vocabulary of key terminology in electronics.

The manual is divided into six sections; DC Circuits, AC Circuits, Semiconductor Devices, Linear Circuits, Digital Circuits, and Soldering Review. Each lab/activity references specific pages in the textbook and care should be taken to point this out to the students. Please note that, with the exception of the DC Circuits and Soldering Review , all sections follow the textbook very closely except for DC Circuits. Even though DC Circuits does not parallel the textbook, it is still organized in a cumulative fashion, so that most activities and labs depend on previous material. Care has been taken to use the same resistance values throughout this DC section.

Each lab covers a single objective or concept. It is constructed in the format:

Objective:
Reference:
Materials Required:
Equipment Required:
Procedure:
Summary:

Where applicable, safety precautions are listed. Each lab is perforated to facilitate ease of student use and teacher grading. Projects are included to help students apply the skills and knowledge acquired. The following areas are covered in the construction of the projects: printed circuit board fabrication, soldering, component identification, use of power and hand tools, screen printing, use of test equipment, circuit analysis and troubleshooting. Printed circuit boards were not included in an attempt to let students try their hand at prototyping and circuit board design.

I would like to thank my daughter, Kimberly for her help in preparing the manuscript in desktop publishing of the manuscript. Thanks also to all the teachers who provided valuable input in my refinement of the lab after using the lab manual in their classes. I would also like to thank the following reviewers for their suggestions:

John Baldwin, South Central Tech College, Faribault, MN
Joe Gryniuk, Lake Washington Technical College, Kirkland, WA
Jim Howe, Lake Washington Technical College, Kirkland, WA
Donald Hofmann, Grayson County College, Denison, TX
Miles Kirkhuff, Lincoln Tech Institute, Allentown, PA
Tim Nichols, Computer Learning Center, Philadelphia, PA
Dan Panetta, Computer Learning Center, Cherry Hill, NJ
Steve Tinker, Education America, Ft. Worth, TX

Appreciation is given to my wife Shirley for her patience and support in the completion of this edition of the lab manual..

Earl D. Gates
2001

LAB 1-1

Fundamentals of Electricity

Objective

The student should be able to identify the terminology associated with electricity in a DC circuit.

Reference

Chapter 1, pages 3-9

Materials Required

None

Equipment Required

None

Notes

Name

Course

Date Due

Definitions

Define the following terms in complete sentences.

1. Atoms
2. Matter
3. Element
4. Compound
5. Proton
6. Electron
7. Valence Shell
8. Ionization
9. Coulomb
10. Difference of Potential
11. Electromotive Force
12. Voltage

Definitions

Define the following terms in complete sentences.

13. Resistance

14. Conductors

15. Insulators

16. Ohm

Questions

Answer the following questions in complete sentences.

1. What letter symbol is used to represent:

    current?

    voltage?

    resistance?

2. What unit is applied to:

    current measurement?

    voltage measurement?

    resistance measurement?

3. What symbol represents the unit of:

    current?

    voltage?

    resistance?

Notes

LAB 1-2

Current

Name

Course

Date Due

Objective

The student should be able to identify the terminology associated with current in a DC circuit.

Reference

Chapter 2, pages 10-16

Materials Required

None

Equipment Required

None

Notes

Definitions

Define the following terms in complete sentences.

1. Coulomb
2. Current
3. Ampere
4. Hole
5. Scientific Notation
6. Milliampere
7. Microampere

Questions

Answer the following questions in complete sentences.

1. What letter symbol is used to represent current?
2. What is the unit of current?
3. The unit of current is represented by what letter symbol?
4. What does I_T represent?

LAB 1-2

Current

Notes

LAB 1-3

Voltage

Name

Course

Date Due

Objective

The student should be able to identify the terminology associated with voltage in a DC circuit.

Reference

Chapter 3, pages 17-30

Materials Required

None

Equipment Required

None

Notes

Definitions

Define the following terms in complete sentences.

1. Difference of Potential
2. Van de Graaf Generator
3. Direct Current
4. Alternating Current
5. Cell
6. Battery
7. Photovoltaic Cell
8. Thermocouple
9. Piezoelectric Effect
10. Primary Cell
11. Secondary Cell
12. Ampere-Hour

Definitions

Define the following terms in complete sentences.

13. Series Cell

14. Series Aiding Cells Configuration

15. Series Opposing Cells Configuration

16. Parallel Cells

17. Series-parallel Cells

18. Voltage Rise

19. Voltage Drop

20. Ground

Questions

Answer the following questions in complete sentences.

1. What letter symbol is used to represent voltage?

2. What unit is applied to voltage measurement?

3. The unit of voltage is represented by what letter symbol?

4. Draw the schematic symbol used to represent:

   AC Generator

   Battery

   Solar Cell

   Thermocouple

   Crystal

5. What does E_T represent?

Notes

LAB 1-4

Resistance

Objective
The student should be able to identify the terminology associated with resistance in a DC circuit.

Reference
Chapter 4, pages 31-48

Materials Required
None

Equipment Required
None

Notes

Name

Course

Date Due

Definitions
Define the following terms in complete sentences.

1. Resistance
2. Resistors
3. Ohm
4. Tolerance
5. Carbon Composition Resistor
6. Wirewound Resistor
7. Film Resistor
8. Potentiometer
9. Rheostat
10. Series Circuit
11. Parallel Circuit
12. Series-parallel Circuit

Definitions

Define the following terms in complete sentences.

13. Ohmmeter

14. Open Circuit

15. Short Circuit

16. Continuity Test

Questions

Answer the following questions in complete sentences.

1. What symbol is used to represent resistance of a resistor?

2. Resistance is represented by what letter symbol?

3. Draw the schematic symbol used to represent:

   fixed resistor.

   variable resistor.

4. Write out the resistor color code.

5. Draw a schematic circuit with resistors and power source connected in:

   series.

   parallel.

   series-parallel.

6. What does R_T represent?

LAB 1-4

Resistance

Notes

LAB 1-5

Unit Conversions

Name

Course

Date Due

Objective

The student should be able to accurately convert from one unit of measure to another.

Reference

Chapter 2, pages 13-15

Materials Required

None

Equipment Required

None

Notes

Directions

Convert the given value to the units indicated. Show all work.

1. 1 A = ____________ mA
2. 2.2 kΩ = ____________ Ω
3. 1.2 MΩ = ____________ kΩ
4. 5,600 Ω = ____________ kΩ
5. 1,000 mW = ____________ W
6. 1.2 mA = ____________ μA
7. 150 mA = ____________ A
8. 10,000 V = ____________ kV
9. 1 MV = ____________ kV
10. 0.47 kΩ = ____________ Ω
11. 0.010 kΩ = ____________ Ω
12. 15.5 mA = ____________ A
13. 575 μA = ____________ A
14. 4.2 MΩ = ____________ Ω
15. 205 μA = ____________ mA

Directions

Convert the given value to the units indicated. Show all work.

16. 2.1 W = ______________________ mW

17. 3 μV = ______________________ V

18. 1.2 MV = ______________________ V

19. 470 Ω = ______________________ kΩ

20. 0.12 kV = ______________________ V

LAB 1-5

Unit Conversions

Notes

LAB 1-6A

Connecting Cells and Batteries

Objective

The student should be able to calculate the total voltage and current available for different series, parallel, and series-parallel cell and battery connections.

Reference

Chapter 3, pages 25-27

Materials Required

None

Equipment Required

None

Notes

Name

Course

Date Due

Directions

Determine the total voltage and current for the cell connections shown. Show all work.

1. All cells are rated 1.5 V at 500 mA.

E_T = ______________ I_T = ______________

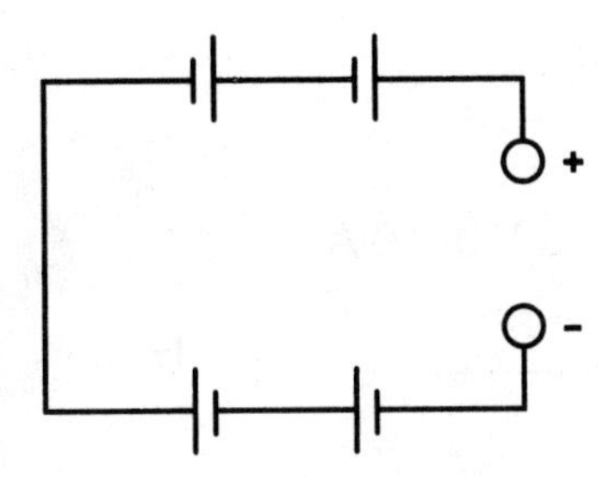

2. All cells are rated 1.5 V at 500 mA.

E_T = ______________ I_T = ______________

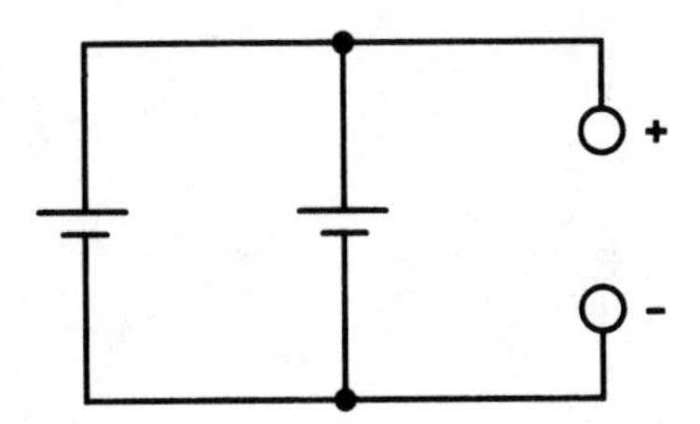

3. All cells are rated 1.2 V at 250 mA.

E_T = ______________ I_T = ______________

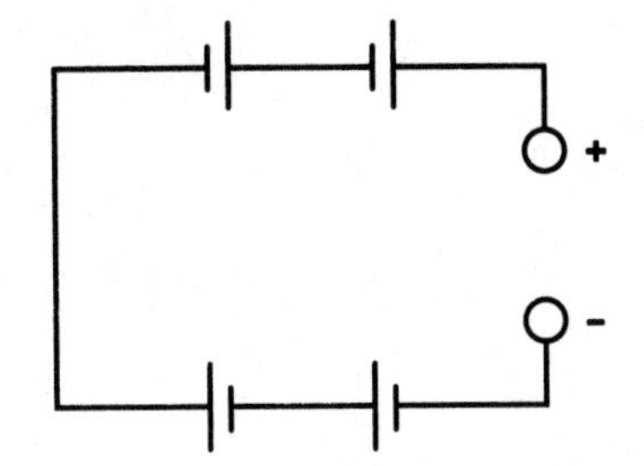

Directions

Determine the total voltage and current for the cell connections shown. Show all work.

4. All cells are rated 1.2 V at 250 mA.

E_T = ______________ I_T = ______________

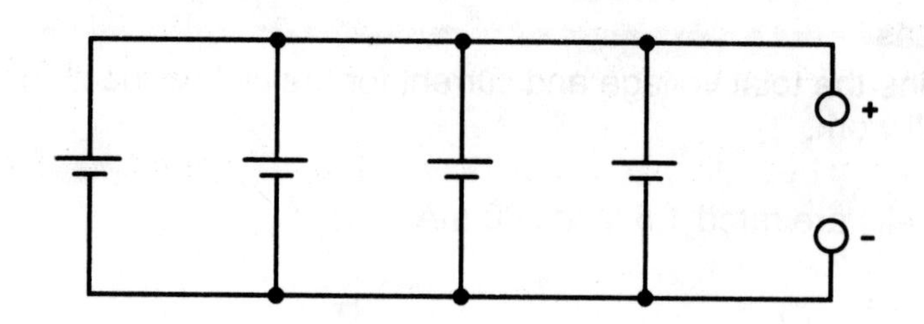

5. All cells are rated 1.5 V at 10 A.

E_T = ______________ I_T = ______________

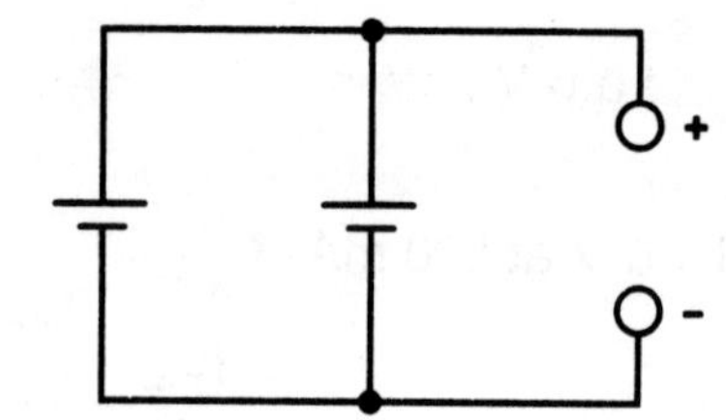

LAB 1-6A

Connecting Cells and Batteries

Notes

LAB 1-6B

Connecting Cells and Batteries

Name
Course
Date Due

Objective

The student should be able to draw schematic circuits of batteries using different series, parallel, and series-parallel cell combinations to obtain desired output voltage and current rating.

Reference

Chapter 3, pages 25-27

Materials Required

None

Equipment Required

None

Notes

Directions

Construct batteries using 1.5 V cells with a current rating of 1 A/h.

1. Draw a schematic of a 3 V battery with a current rating of 1 A/h.

2. Draw a schematic of a 6 V battery with a current rating of 2 A/h.

3. Draw a schematic of a 9 V battery with a current rating of 4 A/h.

Directions

Construct batteries using 1.5 V cells with a current rating of 1 A/h.

4. Draw a schematic of a 12 V battery with a current rating of 6 A/h.

5. Draw a schematic of a 1.5 V battery with a current rating of 10 A/h.

LAB 1-6B

Connecting Cells and Batteries

Notes

LAB 1-7

Measuring Voltage

Name
Course
Date Due

Objective

The student should be able to accurately measure the voltage of individual cells and batteries, cells and batteries in series, and a DC power supply.

Reference

Chapter 6, pages 61-73

Materials Required

AA Dry Cell
C Dry Cell
D Dry Cell
AA Dry Cell
6 V Lantern Battery
9 V Battery
9 V Battery Clip

Equipment Required

DC Analog Voltmeter with Leads
DC Digital Voltmeter with Leads
Adjustable DC Power Supply

Notes

Safety Precautions

Voltmeters are always connected to the circuit or across the voltage source in parallel.

Make sure that the positive lead of the voltage source is always connected to the positive terminal (RED) of the voltmeter and that the negative lead is connected to the negative terminal (BLACK) of the voltmeter.

Procedure

As each step is completed, check it off.

1. Use the digital voltmeter to measure the voltage of each cell and battery indicated on Table 1-7-1.
2. Record the voltage on Table 1-7-1 under the column marked Digital Voltmeter.
3. Use the analog voltmeter to measure the voltage for each cell and battery indicated on Table 1-7-1.
4. Record the voltage on Table 1-7-1 under the column marked Analog Voltmeter.
5. Place the negative terminal of the C-cell on the positive terminal of the D-Cell.
6. Use the voltmeter of your choice to measure the voltage. Measure and record the voltage across the battery combination and record the results on Table 1-7-2 under the appropriate voltmeter.
7. Connect the AA-cell, C-cell, and D-cell in series. The cells can be held together by masking tape.
8. Use the voltmeter of your choice to measure the voltage. Measure and record the voltage across the combination and record the results on Table 1-7-2 under the appropriate voltmeter.

Procedure

As each step is completed, check it off. Use the adjustable power supply for steps 9 - 17.

9. Connect the analog voltmeter to the output terminals and record the voltage readings for steps 10 - 14 on Table 1-7-3 under the column marked Analog Voltmeter.
10. Initially set the control knob to the minimum output voltage position, fully counterclockwise, and record the voltage.
11. Set the voltage adjustment knob to the 9 o'clock position and record the voltage.
12. Set the voltage adjustment knob to the 12 o'clock position and record the voltage.
13. Set the voltage adjustment knob to the 3 o'clock position and record the voltage.
14. Set the voltage adjustment knob to the maximum position, fully clockwise, and record the voltage.
15. Connect the digital voltmeter to the output terminals and record the voltage readings for step 16 on Table 1-7-3 under the column marked Digital Voltmeter.
16. Repeat steps 10 - 14 using the digital voltmeter.
17. Disconnect the voltmeter and turn off the power supply.

Summary

Most cells produce approximately 1.5 V. When the cell is fresh, the voltage is slightly higher. As the cell is used, the voltage produced decreases to less than 1.5 V. When the cells are connected in series, the voltage increases. Batteries are combinations of cells, the voltage produced is multiples of 1.5 V. A 6-V lantern battery is constructed of four 1.5-V cells connected in series. A 9-V battery has six 1.5-V cells interconnected in series.

A DC power supply produces a DC voltage, like the cells and batteries. The advantage of the power supply is the voltage does not decrease with usage. An adjustable power supply is handy because many different voltages may be generated from the same unit.

When using a digital voltmeter the reading is exact. However, with an analog voltmeter the reading has to be interpreted.

Notes

LAB 1-7

Measuring Voltage

Name
Course
Date Due

Notes

Directions

Complete Tables 1-7-1 and 1-7-2 using the steps on page 15.

FIGURE 1-7-1 • DRY CELLS

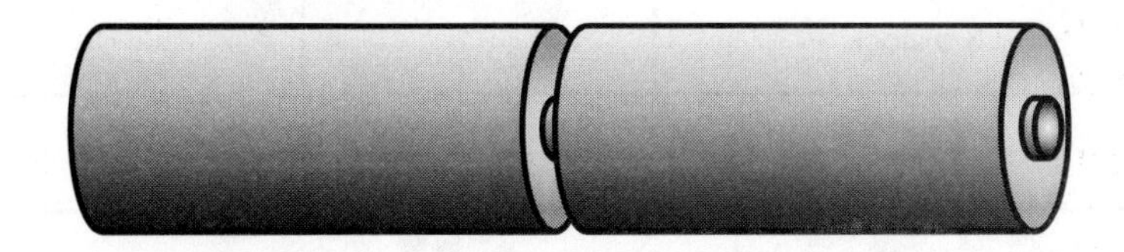

TABLE 1-7-1

Cell/Battery Type	Digital Voltmeter	Analog Voltmeter
AA		
C		
D		
6 V		
9 V		

TABLE 1-7-2

Cell/Battery Type	Digital Voltmeter	Analog Voltmeter
C-cell + D-cell		
AA-cell + C-cell + D-cell		

Directions

Complete Table 1-7-3 using the steps on page 16.

TABLE 1-7-3

Position	Digital Voltmeter	Analog Voltmeter
Minimum		
9 o'clock		
12 o'clock		
3 o'clock		
Maximum		

Notes

LAB 1-8

Resistor Color Code

Name	
Course	
Date Due	

Objective

The student should be able to encode and decode given resistor values using the resistor color code.

Reference

Chapter 4, pages 37-40

Materials Required

None

Equipment Required

None

Notes

Directions

Identify the resistance value and tolerance for each of the color codes listed.

	Value	Tolerance
1. Yellow, Violet, Red, Silver	__________	__________
2. Red, Red, Yellow, Gold	__________	__________
3. Brown, Black, Brown, None	__________	__________
4. Brown, Black, Red, Silver	__________	__________
5. Yellow, Violet, Orange, Gold	__________	__________
6. Red, Violet, Green, None	__________	__________
7. Brown, Black, Yellow, Gold	__________	__________
8. Red, Gray, Yellow, None	__________	__________
9. Orange, White, Brown, Silver	__________	__________
10. Blue, Green, Black, Gold	__________	__________
11. Orange, Orange, Orange, None	__________	__________
12. Red, Red, Red, Silver	__________	__________
13. Brown, Black, Black, Gold	__________	__________
14. Blue, Green, Orange, Silver	__________	__________

Directions

Write the color code for the following resistor values.

	1st Band	2nd Band	3rd Band	Tolerance
1. 2200 Ω ± 10%				
2. 10 Ω ± 5%				
3. 10,000,000 Ω ± 20%				
4. 470 Ω ± 10%				
5. 390,000 Ω ± 5%				
6. 68,000 Ω ± 20%				
7. 27 Ω ± 5%				
8. 4,700,000 Ω ± 10%				
9. 10,000 Ω ± 20%				
10. 56 Ω ± 10%				
11. 220 Ω ± 5%				
12. 5600 Ω ± 20%				
13. 1000 Ω ± 5%				
14. 470,000 Ω ± 10%				
15. 390 Ω ± 20%				
16. 270 Ω ± 10%				
17. 330,000 Ω ± 5%				
18. 6800 Ω ± 20%				

LAB 1-8

Resistor Color Code

Notes

LAB 1-9

Resistors in Series

Name

Course

Date Due

Objective

The student should be able to compute total resistance for circuits with resistors in series.

Reference

Chapter 4, pages 40-41

Materials Required

None

Equipment Required

Calculator

Notes

Directions

Determine the total resistance for each of the following circuits. Show all work. The formula used to compute the total series resistance is:

$$R_T = R_1 + R_2 + R_3 + \ldots R_n$$

1. R_T = ____________________

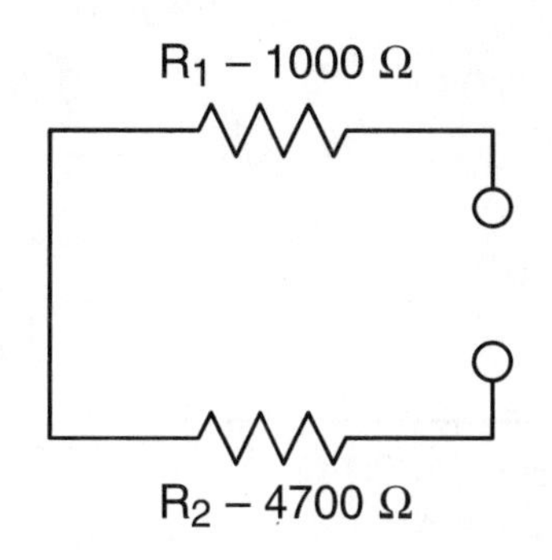

2. R_T = ____________________

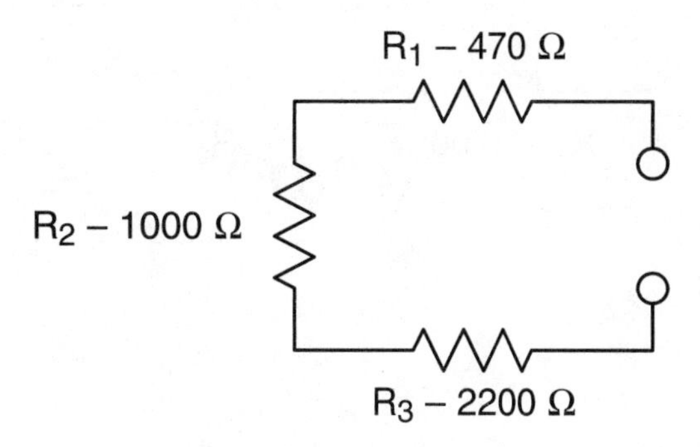

3. R_T = ____________________

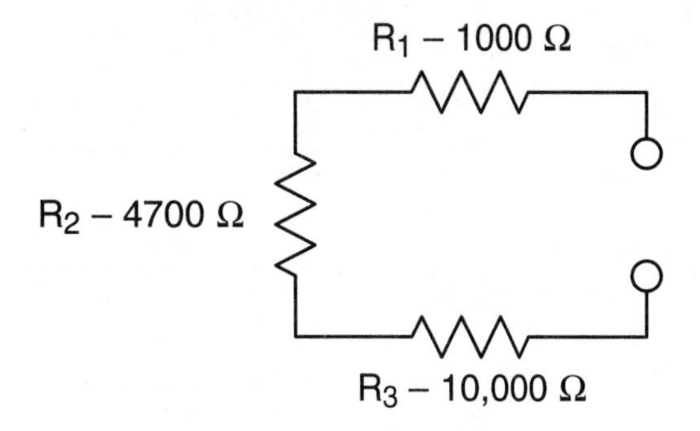

Directions

Determine the total resistance for each of the following circuits. Show all work. The formula used to compute the total series resistance is:

$$R_T = R_1 + R_2 + R_3 + \dots R_n$$

4. R_T =

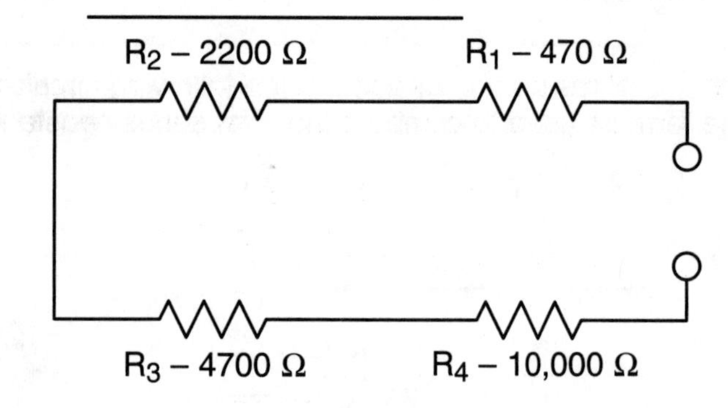

5. R_T =

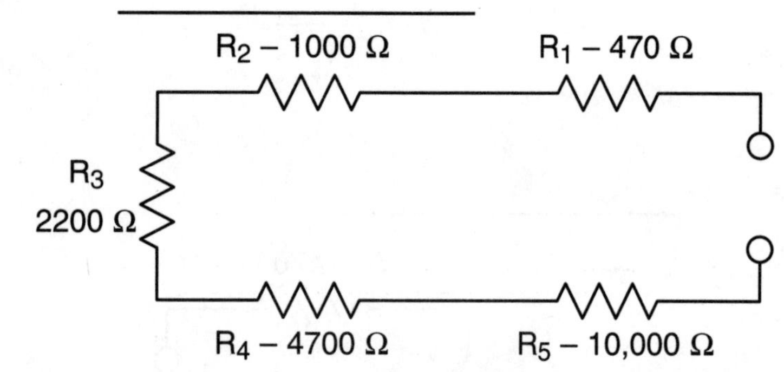

LAB 1-9

Resistors in Series

Notes

LAB 1-10

Resistors in Parallel

Name

Course

Date Due

Objective

The student should be able to compute the total resistance for circuits with resistors in parallel.

Reference

Chapter 4, pages 41-43

Materials Required

None

Equipment Required

Calculator

Notes

Directions

Determine the total resistance for each of the following circuits. Show all work. The formula used to compute the total parallel resistance is:

$$\frac{1}{R_T} = \frac{1}{R_1} + \frac{1}{R_2} + \frac{1}{R_3} + \ldots\frac{1}{R_n}$$

1. R_T = ______________

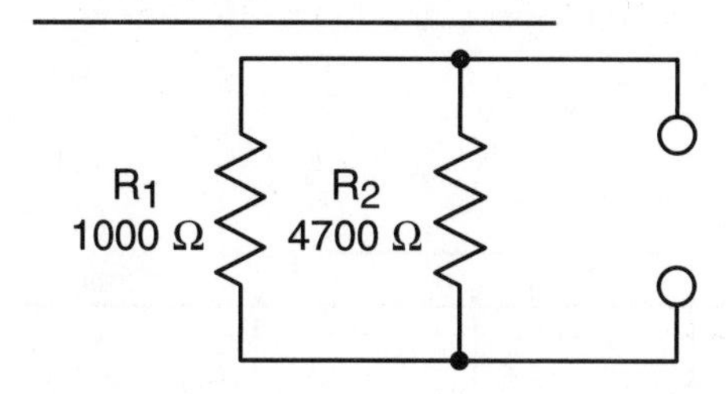

2. R_T = ______________

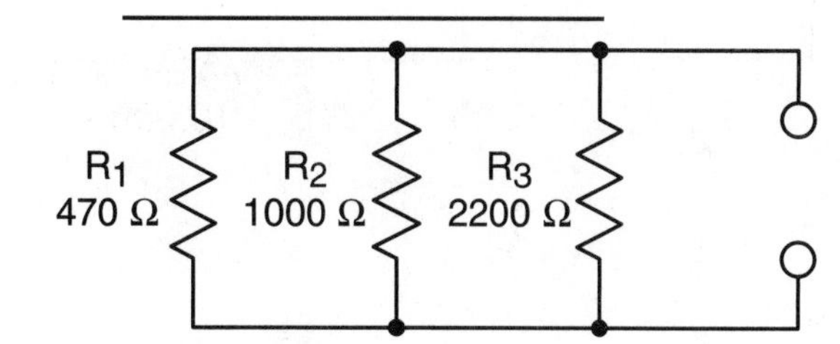

3. R_T = ______________

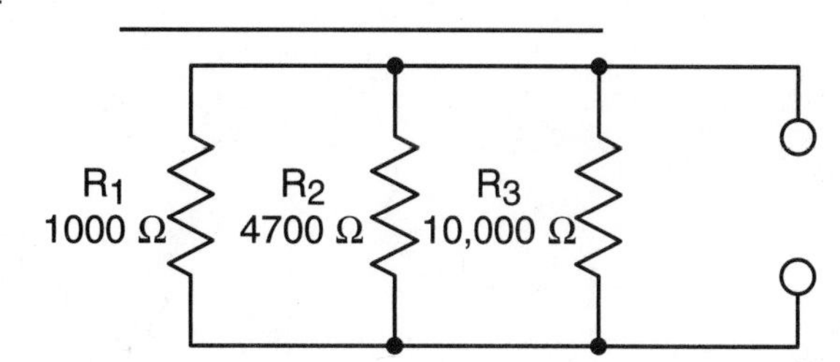

Directions

Determine the total resistance for each of the following circuits. Show all work. The formula used to compute the total parallel resistance is:

$$\frac{1}{R_T} = \frac{1}{R_1} + \frac{1}{R_2} + \frac{1}{R_3} + \ldots \frac{1}{R_n}$$

4. $R_T =$ ______________________

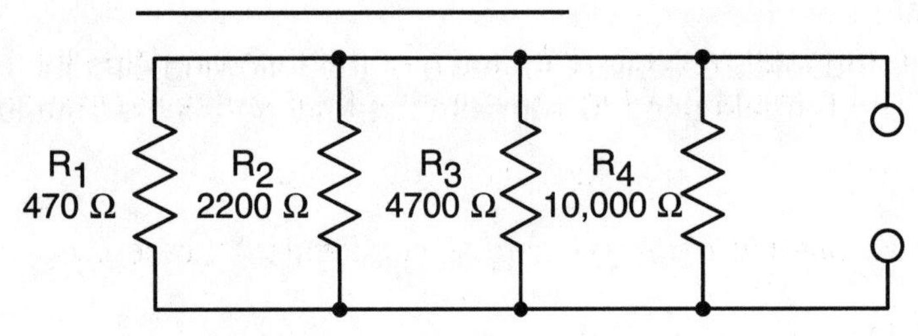

5. $R_T =$ ______________________

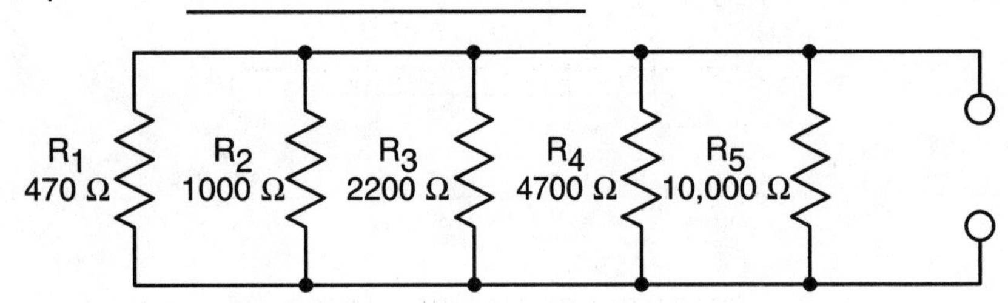

LAB 1-10

Resistors in Parallel

Notes

LAB 1-11

Resistors in Series-Parallel

Name	
Course	
Date Due	

Objective

The student should be able to compute the total resistance for resistors in a series-parallel circuit.

Reference

Chapter 4, pages 43-46

Materials Required

None

Equipment Required

Calculator

Notes

Directions

Determine the total resistance for each of the following circuits. Show all work. The formula used to compute the total series resistance is:

$$R_T = R_1 + R_2 + R_3 + \ldots R_n$$

The formula used to compute the total parallel resistance is:

$$\frac{1}{R_T} = \frac{1}{R_1} + \frac{1}{R_2} + \frac{1}{R_3} + \ldots \frac{1}{R_n}$$

1. R_T = ______________

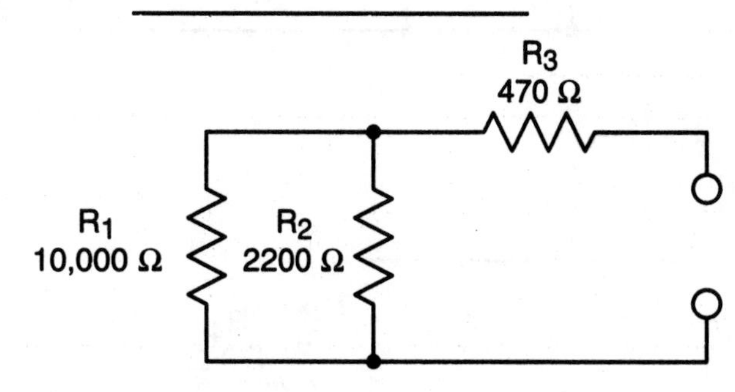

2. R_T = ______________

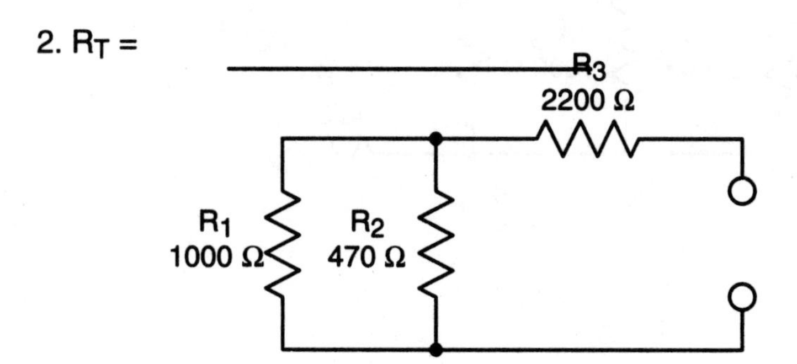

3. R_T = ______________

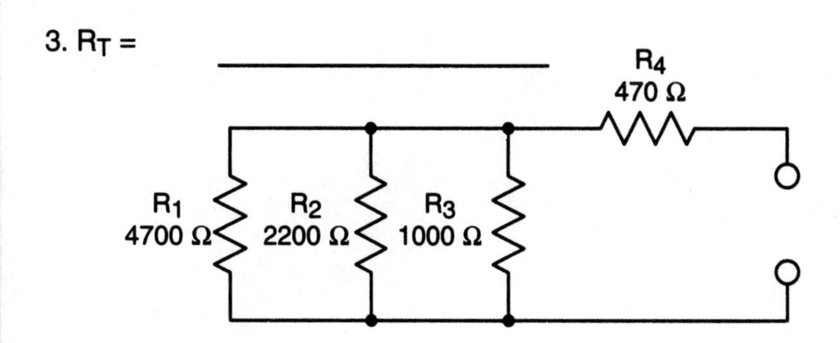

Directions

Determine the total resistance for each of the following circuits. Show all work. The formula used to compute the total series resistance is:

$$R_T = R_1 + R_2 + R_3 + \ldots R_n$$

The formula used to compute the total parallel resistance is:

$$\frac{1}{R_T} = \frac{1}{R_1} + \frac{1}{R_2} + \frac{1}{R_3} + \ldots \frac{1}{R_n}$$

4. R_T = ____________________

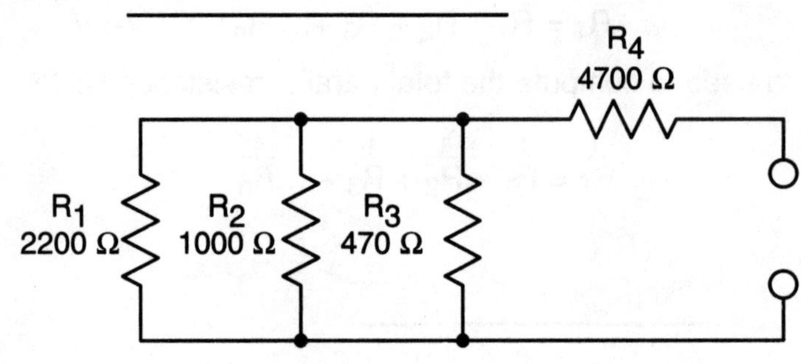

5. R_T = ____________________

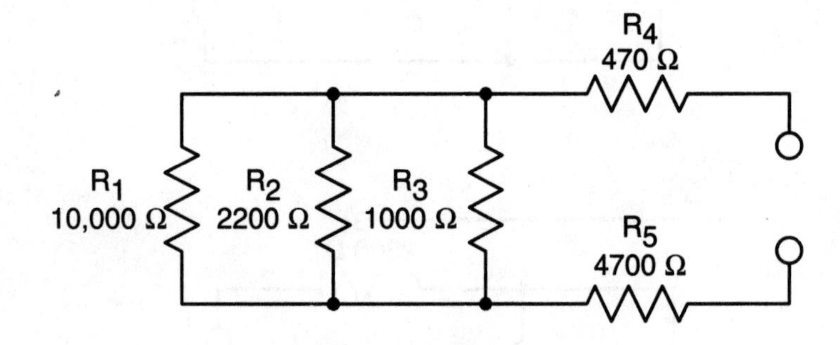

LAB 1-11

Resistors in Series-Parallel

Notes

LAB 1-12

Resistor Measurements

Name

Course

Date Due

Objective

The student should be able to accurately read resistor color codes and verify the results with an ohmmeter.

Reference

Chapter 6, pages 61-73

Materials Required

1/2 W resistors:

Yellow - Violet - Brown
Brown - Black - Red
Red - Red - Red
Yellow - Violet - Red
Brown - Black - Orange

Note: The tolerance band for the resistors may be gold, silver, or none.

Equipment Required

Ohmmeter with Leads

Notes

Procedure

As each step is completed, check it off.

1. Obtain one of each of the resistors indicated in the Materials Required list.
2. Lay out the resistor on the bench in the order listed on Table 1-12-1.
3. Using the resistor color code, determine the value of each of the resistors, including the tolerance. Record the value on Table 1-12-1 under the column Color Code Value.
4. Determine the tolerance value of each resistor using the resistor color code. Record the value on Table 1-12-1 under the column Tolerance Value.
5. Using the tolerance value of each resistor, calculate the minimum and maximum tolerance range that each resistor can vary.
6. Record the minimum and maximum tolerance range on Table 1-12-1 under the column Tolerance Range.
7. Using the ohmmeter, measure the value of each resistor.
8. Record the actual value of each resistor on Table 1-12-1 under the column Actual Value.
9. Compare the actual value of each resistor with the minimum and maximum calculated values.
10. Identify whether the resistor is in tolerance with a Yes or No under the column identified In Tolerance.

Directions

Complete Table 1-12-1 using the steps on page 27.

TABLE 1-12-1

Resistor	Color Code Value	Tolerance Value	Tolerance Range		Actual Value	In Tolerance
			Maximum	Minimum		
R_1 - Yel, Vio, Brn						
R_2 - Brn, Blk, Red						
R_3 - Red, Red, Red						
R_4 - Yel, Vio, Red						
R_5 - Brn, Blk, Org						

Summary

The resistor color code is the most common method of identifying carbon composition resistors. Carbon composition resistors are the most commonly used resistors for projects.

The tolerance band gives an indication of how much a resistor can vary and still be acceptable for the value indicated. If the resistor varies beyond tolerance range, it should be considered unacceptable. In some circuits it is very important that the resistor be within the range specified, otherwise the circuit may not work.

Do not dispose of these resistors. They will be used for other labs in the manual.

Notes

LAB 1-13

Series Resistor Measurements

Name
Course
Date Due

Objective

The student should be able to accurately measure resistors placed in series and calculate the total resistance in series as a check.

Reference

Chapter 6, pages 61-73

Materials Required

Resistors

R_1 - 470 Ω, 1/2 W,
Yellow - Violet - Brown

R_2 - 1000 Ω, 1/2 W,
Brown - Black - Red

R_3 - 2200 Ω, 1/2 W,
Red - Red - Red

R_4 - 4700 Ω, 1/2 W,
Yellow - Violet - Red

R_5 - 10,000 Ω, 1/2 W,
Brown - Black - Orange

NOTE*: The tolerance band for the resistors may be gold, silver, or none.*

Equipment Required

Ohmmeter with Leads

Notes

Procedure

As each step is completed, check it off.

1. Use the same resistors that were used for Lab 1-12.
2. Transfer the actual values from Lab 1-12, Table 1-12-1 to Table 1-13-1 under the column Actual Value.
3. Connect the resistors in series as indicated on Table 1-13-1. The symbol "+" indicates the resistors are to be connected in series.
4. Measure the total resistance using the ohmmeter.
5. Record the total resistance measured under the column marked Total Resistance Measured on Table 1-13-1.
6. Repeat steps 3 - 5 for each of the series resistor combinations listed on Table 1-13-1.
7. Transfer the calculated resistance values from Lab 1-9 to Table 1-13-1 under the column marked Total Calculated Resistance. Carefully check the values of each resistor combination when making this transfer.
8. On Table 1-13-1, compare the total resistance measured with the calculated resistance.
9. Indicate whether the calculated value is within 50 Ω of the actual value with a Yes or No under the column marked Total Resistance Comparison.

LAB 1-13

Series Resistor Measurements

Directions

Complete Table 1-13-1 using the steps on page 29.

TABLE 1-13-1

Resistor Combination		Color Code Value	Actual Value	Total Resistance		Total Resistance Comparison
				Measured	Calculated	
1. $R_2 + R_4$	R_2					
	R_4					
2. $R_1 + R_2 + R_3$	R_1					
	R_2					
	R_3					
3. $R_2 + R_4 + R_5$	R_2					
	R_4					
	R_5					
4. $R_1 + R_3 + R_4 + R_5$	R_1					
	R_3					
	R_4					
	R_5					
5. $R_1 + R_2 + R_3 + R_4 + R_5$	R_1					
	R_2					
	R_3					
	R_4					
	R_5					

Summary

In a series circuit the total resistance is the sum of the individual resistors. If the value measured does not agree with the calculated value, recheck your calculations. If the answer still disagrees, remeasure the circuit. Check for faulty connections at each of the resistor leads in the circuit, poor connections of test leads, or an improperly zeroed ohmmeter.

Notes

LAB 1-14

Parallel Resistor Measurements

Name

Course

Date Due

Objective

The student should be able to accurately measure resistors placed in parallel and calculate the total resistance in parallel as a check.

Reference

Chapter 6, pages 61-73

Materials Required

Resistors

R_1 - 470 Ω, 1/2 W, Yellow - Violet - Brown

R_2 - 1000 Ω, 1/2 W, Brown - Black - Red

R_3 - 2200 Ω, 1/2 W, Red - Red - Red

R_4 - 4700 Ω, 1/2 W, Yellow - Violet - Red

R_5 - 10,000 Ω, 1/2 W, Brown - Black - Orange

***NOTE**: The tolerance band for the resistors may be gold, silver, or none.*

Equipment Required

Ohmmeter with Leads

Notes

Procedure

As each step is completed, check it off.

1. Use the same resistors that were used for Lab 1-12.
2. Transfer the actual values from Lab 1-12, Table 1-12-1 to Table 1-14-1 under the column Actual Value.
3. Connect the resistors in parallel as indicated on Table 1-14-1. The symbol "II" indicates the resistors are to be connected in parallel.
4. Measure the total resistance using the ohmmeter.
5. Record the total resistance measured under the column marked Total Resistance Measured on Table 1-14-1.
6. Repeat steps 3 - 5 for each of the series resistor combinations listed on Table 1-14-1.
7. Transfer the calculated resistance values from Lab 1-10 to Table 1-14-1 under the column marked Total Calculated Resistance. Carefully check the values of each resistor combination when making this transfer.
8. On Table 1-14-1, compare the total resistance measured with the calculated resistance.
9. Indicate whether the calculated value is within 50 Ω of the actual value with a Yes or No under the column marked Total Resistance Comparison.

Directions

Complete Table 1-14-1 using the steps on page 31.

TABLE 1-14-1

Resistor Combination		Color Code Value	Actual Value	Total Resistance		Total Resistance Comparison
				Measured	Calculated	
1. $R_2 \parallel R_4$	R_2					
	R_4					
2. $R_1 \parallel R_2 \parallel R_3$	R_1					
	R_2					
	R_3					
3. $R_2 \parallel R_4 \parallel R_5$	R_2					
	R_4					
	R_5					
4. $R_1 \parallel R_3 \parallel R_4 \parallel R_5$	R_1					
	R_3					
	R_4					
	R_5					
5. $R_1 \parallel R_2 \parallel R_3 \parallel R_4 \parallel R_5$	R_1					
	R_2					
	R_3					
	R_4					
	R_5					

Summary

In a parallel circuit the total resistance is always smaller than the smallest resistors. If the value measured does not agree with the calculated value, recheck your calculations. If the answer still disagrees, remeasure the circuit. Check for faulty connections at each of the resistor leads in the circuit, poor connections of test leads, or an improperly zeroed ohmmeter.

LAB 1-14

Parallel Resistor Measurements

Notes

LAB 1-15

Series-Parallel Resistor Measurements

Name

Course

Date Due

Objective

The student should be able to accurately measure resistors connected in a series-parallel combination and calculate the total resistance of a series-parallel combination.

Reference

Chapter 6, pages 61-73

Materials Required

Resistors

- R_1 - 470 Ω, 1/2 W, Yellow - Violet - Brown
- R_2 - 1000 Ω, 1/2 W, Brown - Black - Red
- R_3 - 2200 Ω, 1/2 W, Red - Red - Red
- R_4 - 4700 Ω, 1/2 W, Yellow - Violet - Red
- R_5 - 10,000 Ω, 1/2 W, Brown - Black - Orange

***NOTE**: The tolerance band for the resistors may be gold, silver, or none.*

Equipment Required

Ohmmeter with Leads

Notes

Procedure

As each step is completed, check it off.

1. Use the same resistors that were used for Lab 1-12.
2. Transfer the actual values from Lab 1-12, Table 1-12-1 to Table 1-15-1 under the column Actual Value.
3. Connect the resistors in parallel as indicated on Table 1-15-1. The symbol "+" indicates the resistors are to be connected in series and the symbol "||" indicates the resistors are to be connected in parallel.
4. Measure the total resistance using the ohmmeter.
5. Record the total resistance measured under the column marked Total Resistance Measured on Table 1-15-1.
6. Repeat steps 3 - 5 for each of the series-parallel resistor combinations listed on Table 1-15-1.
7. Transfer the calculated resistance values from Lab 1-11 to Table 1-15-1 under the column marked Total Calculated Resistance. Carefully check the values of each resistor combination when making this transfer.
8. On Table 1-15-1, compare the total resistance measured with the calculated resistance.
9. Indicate whether the calculated value is within 50 Ω of the actual value with a Yes or No under the column marked Total Resistance Comparison.

Directions

Complete Table 1-15-1 using the steps on page 33.

TABLE 1-15-1

Resistor Combination		Color Code Value	Actual Value	Total Resistance		Total Resistance Comparison
				Measured	Calculated	
1. $R_1 + R_3 \parallel R_4$	R_1					
	R_3					
	R_4					
2. $R_1 \parallel R_2 + R_3$	R_1					
	R_2					
	R_3					
3. $R_1 + R_2 \parallel R_3 \parallel R_4$	R_1					
	R_2					
	R_3					
	R_4					
4. $R_1 \parallel R_2 \parallel R_3 + R_4$	R_1					
	R_2					
	R_3					
	R_4					
5. $R_1 + R_2 \parallel R_3 \parallel R_5 + R_4$	R_1					
	R_2					
	R_3					
	R_4					
	R_5					

Summary

Prior to doing any measurements or calculations, the circuit to be connected should be drawn out first. When doing the calculations for a series-parallel circuit, the resistance of the parallel components must be found first. When doing the calculations, it helps to redraw the circuit each time. If the value measured does not agree with the calculated value, recheck your calculations. If the answer still disagrees, remeasure the circuit. Check for faulty connections at each of the resistor leads in the circuit, poor connections of test leads, or an improper zeroed ohmmeter.

LAB 1-15

Series-Parallel Resistor Measurements

Notes

LAB 1-16

Ohm's Law

Name

Course

Date Due

Objective

The student should be able to use Ohm's Law for solving problems.

Reference

Chapter 5, pages 49-60

Materials Required

None

Equipment Required

Calculator

Notes

Directions

Solve the following problems using Ohm's Law. Show all work.

1. $I = 100$ mA $E = ?$ $R = 100\ \Omega$ $E =$ ____________

2. $I = 2$ A $E = ?$ $R = 50\ \Omega$ $E =$ ____________

3. $I = ?$ $E = 250$ V $R = 500\ \Omega$ $I =$ ____________

4. $I = ?$ $E = 50$ V $R = 10\ k\Omega$ $I =$ ____________

5. $I = 2$ A $E = 100$ V $R = ?$ $R =$ ____________

Directions

Solve the following problems using Ohm's Law. Show all work.

6. I = 10 mA   E = 50 V   R = ?   R = ________

7. I = 0.01 A   E = ?   R = 100 Ω   E = ________

8. I = 2 A   E = 10 V   R = ?   R = ________

9. I = ?   E = 120 V   R = 100 kΩ   I = ________

10. I = 10 A   E = ?   R = 120 Ω   E = ________

LAB 1-16

Ohm's Law

Notes

LAB 1-17

Power

Name

Course

Date Due

Objective

The student should be able to solve problems involving power, current, and voltage.

Reference

Chapter 7, pages 74-78

Materials Required

None

Equipment Required

Calculator

Notes

Directions

Solve the following problems using Watt's Law. Show all work.

1. P = 1 W   I = ?   E = 10 V   I = ________

2. P = 0.5 W   I = ?   E = 100 V   I = ________

3. P = ?   I = 100 mA   E = 50 V   P = ________

4. P = ?   I = 400 mA   E = 200 V   P = ________

5. P = 2 W   I = 100 A   E = ?   E = ________

Directions

Solve the following problems using Watt's Law. Show all work.

6. P = 0.5 W   I = 28 mA   E = ?   E = __________

7. P = 1 W   I = ?   E = 100 V   I = __________

8. P = 4 V   I = 100 mA   E = ?   E = __________

9. P = ?   I = 90 mA   E = 30 V   P = __________

10. P = 10 W   I = ?   E = 120 V   I = __________

LAB 1-17

Power

Notes

LAB 1-18

Series DC Circuits

Name

Course

Date Due

Objective

The student should be able to solve for all unknown values in a series DC circuit.

Reference

Chapter 8, pages 79-93

Materials Required

None

Equipment Required

Calculator

Notes

Directions

Solve for all values listed for each problem shown. Show all work and draw all equivalent circuits.

Ohm's Law: $I = E / R$

Formulas for series circuit:

$R_T = R_1 + R_2 + R_3 + \ldots R_n$

$E_T = E_1 + E_2 + E_3 + \ldots E_n$

$I_T = I_1 = I_2 = I_3 = \ldots I_n$

$P_T = P_1 + P_2 + P_3 + \ldots P_n$

1.

	R	I	E	P
R_1	1000 Ω			
R_2	4700 Ω			
Total			20 V	

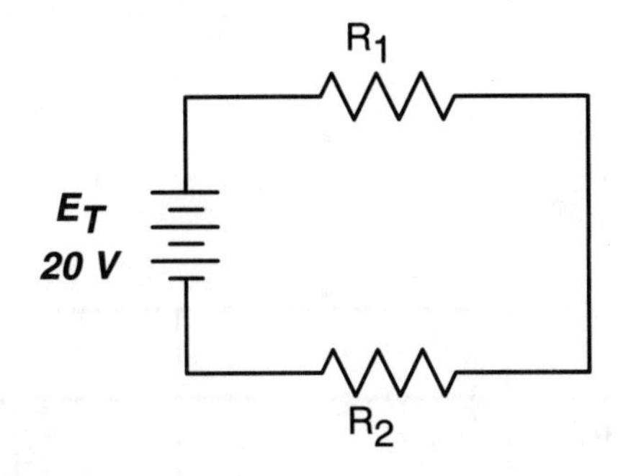

2.

	R	I	E	P
R_1	470 Ω			
R_2	1000 Ω			
R_3	2200 Ω			
Total			20 V	

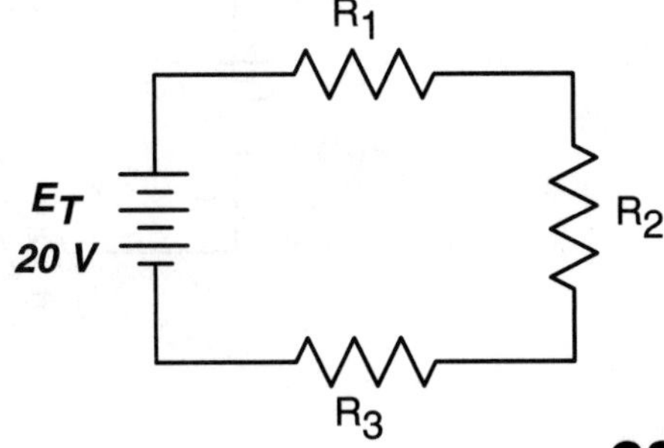

Directions

Solve for all values listed for each problem shown. Show all work and draw all equivalent circuits.

LAB 1-18

Series DC Circuits

Ohm's Law: $I = E / R$ Formulas for series circuit: $R_T = R_1 + R_2 + R_3 + \ldots R_n$

$E_T = E_1 + E_2 + E_3 + \ldots E_n$

$I_T = I_1 = I_2 = I_3 = \ldots I_n$

$P_T = P_1 + P_2 + P_3 + \ldots P_n$

Notes

3.

	R	I	E	P
R_1	1000 Ω			
R_2	4700 Ω			
R_3	10,000 Ω			
Total			*20 V*	

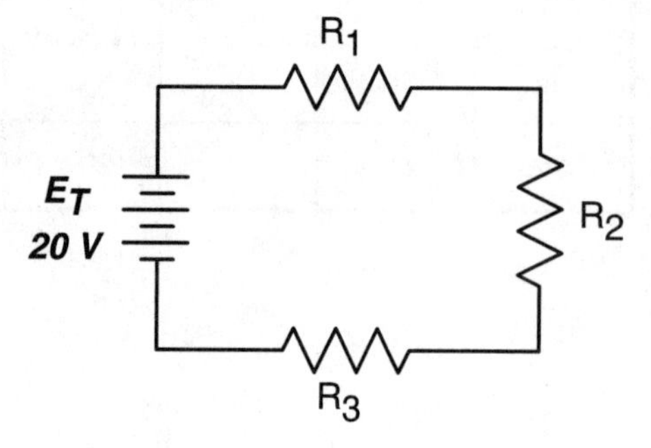

4.

	R	I	E	P
R_1	470 Ω			
R_2	2200 Ω			
R_3	4700 Ω			
R_4	10,000 Ω			
Total			20 V	

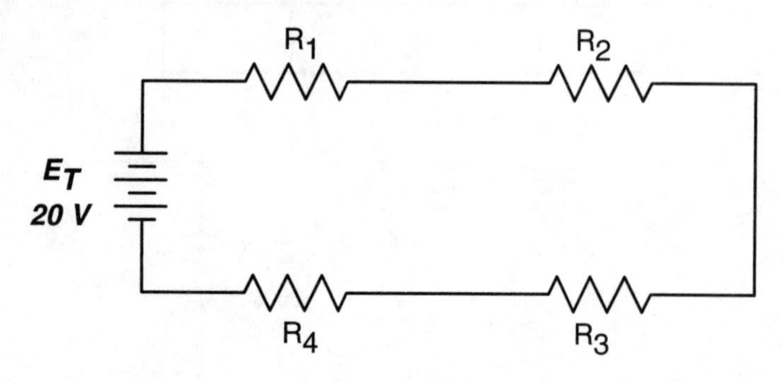

LAB 1-18

Series DC Circuits

Name

Course

Date Due

Directions

Solve for all values listed for each problem shown. Show all work and draw all equivalent circuits.

Ohm's Law: $I = E / R$

Formulas for series circuit:

$R_T = R_1 + R_2 + R_3 + \dots R_n$

$E_T = E_1 + E_2 + E_3 + \dots E_n$

$I_T = I_1 = I_2 = I_3 = \dots I_n$

$P_T = P_1 + P_2 + P_3 + \dots P_n$

5.

	R	I	E	P
R_1	470 Ω			
R_2	1000 Ω			
R_3	2200 Ω			
R_4	4700 Ω			
R_5	10,000 Ω			
Total			20 V	

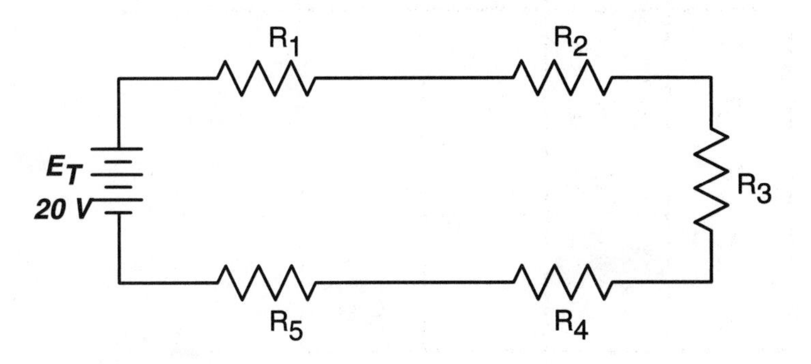

Notes

LAB 1-19

Parallel DC Circuits

Name
Course
Date Due

Objective

The student should be able to solve for all unknown values in a parallel DC circuit.

Reference

Chapter 8, pages 79-93

Materials Required

None

Equipment Required

Calculator

Notes

Directions

Solve for all values listed for each problem shown. Show all work and draw all equivalent circuits.

Ohm's Law: $I = E / R$

Formulas for parallel circuit:

$$\frac{1}{R_T} = \frac{1}{R_1} + \frac{1}{R_2} + \frac{1}{R_3} + \ldots \frac{1}{R_n}$$

$$E_T = E_1 = E_2 = E_3 = \ldots E_n$$

$$I_T = I_1 + I_2 + I_3 + \ldots I_n$$

$$P_T = P_1 + P_2 + P_3 + \ldots P_n$$

1.

	R	I	E	P
R_1	1000 Ω			
R_2	4700 Ω			
Total			20 V	

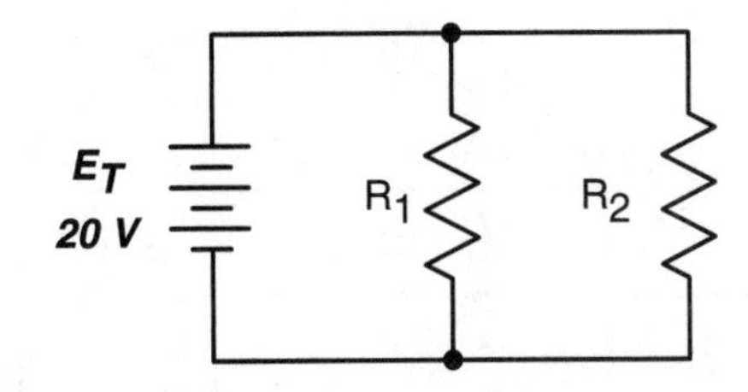

2.

	R	I	E	P
R_1	470 Ω			
R_2	1000 Ω			
R_3	2200 Ω			
Total			20 V	

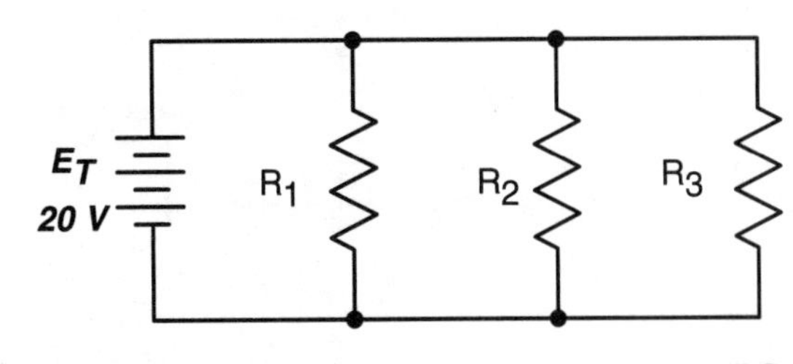

Directions

Solve for all values listed for each problem shown. Show all work and draw all equivalent circuits.

Ohm's Law: $I = E / R$ Formulas for parallel circuit:

$$\frac{1}{R_T} = \frac{1}{R_1} + \frac{1}{R_2} + \frac{1}{R_3} + \ldots \frac{1}{R_n}$$

$$E_T = E_1 = E_2 = E_3 = \ldots E_n$$

$$I_T = I_1 + I_2 + I_3 + \ldots I_n$$

$$P_T = P_1 + P_2 + P_3 + \ldots P_n$$

3.

	R	I	E	P
R_1	1000 Ω			
R_2	4700 Ω			
R_3	10,000 Ω			
Total			20 V	

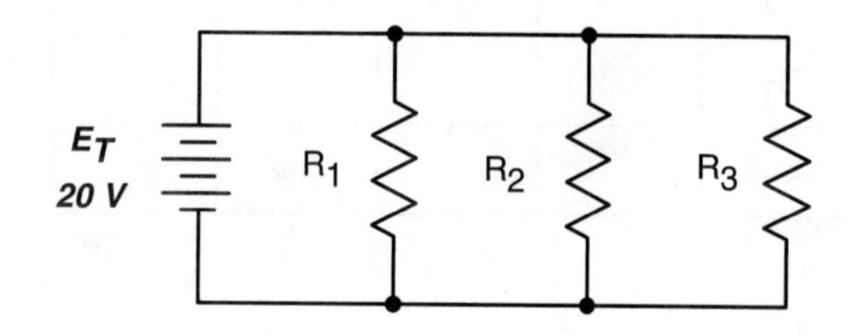

4.

	R	I	E	P
R_1	470 Ω			
R_2	2200 Ω			
R_3	4700 Ω			
R_4	10,000 Ω			
Total			20 V	

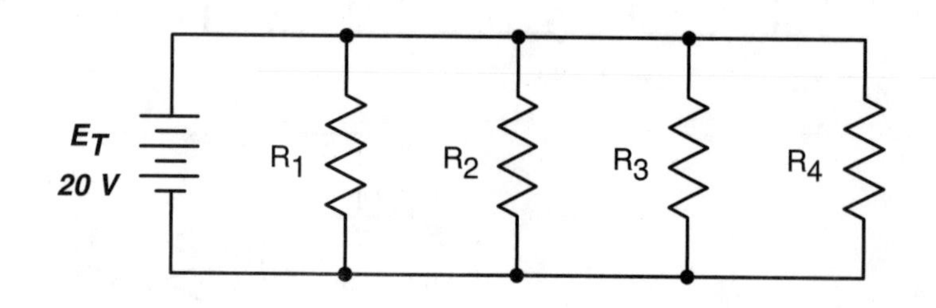

LAB 1-19

Parallel DC Circuits

Notes

LAB 1-19

Parallel DC Circuits

Name
Course
Date Due

Directions

Solve for all values listed for each problem shown. Show all work and draw all equivalent circuits.

Ohm's Law: $I = E / R$ Formulas for parallel circuit:

$$\frac{1}{R_T} = \frac{1}{R_1} + \frac{1}{R_2} + \frac{1}{R_3} + \ldots \frac{1}{R_n}$$

$$E_T = E_1 = E_2 = E_3 = \ldots E_n$$

$$I_T = I_1 + I_2 + I_3 + \ldots I_n$$

$$P_T = P_1 + P_2 + P_3 + \ldots P_n$$

5.

	R	I	E	P
R_1	470 Ω			
R_2	1000 Ω			
R_3	2200 Ω			
R_4	4700 Ω			
R_5	10,000 Ω			
Total			20 V	

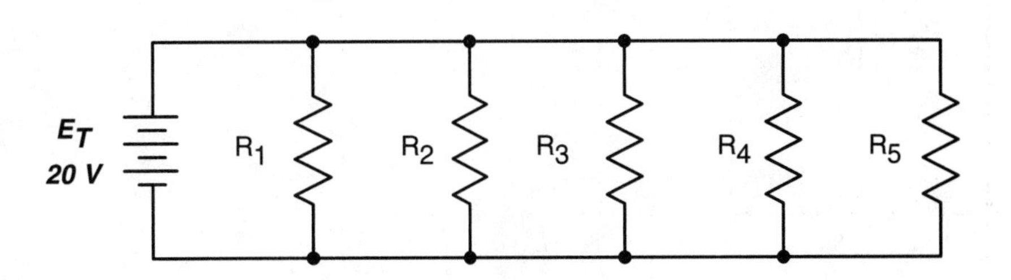

LAB 1-19

Parallel DC Circuits

Notes

LAB 1-20

Series-Parallel DC Circuits

Name
Course
Date Due

Objective

The student should be able to solve for all unknown values in a series-parallel DC circuit.

Reference

Chapter 8, pages 79-93

Materials Required

None

Equipment Required

Calculator

Notes

Directions

Solve for all values listed for each problem shown. Show all work and draw all equivalent circuits.

Ohm's Law: $I = E / R$

Formulas for series circuit:

$R_T = R_1 + R_2 + R_3 + \dots R_n$

$E_T = E_1 + E_2 + E_3 + \dots E_n$

$I_T = I_1 = I_2 = I_3 = \dots I_n$

$P_T = P_1 + P_2 + P_3 + \dots P_n$

Formulas for parallel circuit:

$\frac{1}{R_T} = \frac{1}{R_1} + \frac{1}{R_2} + \frac{1}{R_3} + \dots \frac{1}{R_n}$

$E_T = E_1 = E_2 = E_3 = \dots E_n$

$I_T = I_1 + I_2 + I_3 + \dots I_n$

$P_T = P_1 + P_2 + P_3 + \dots P_n$

1.

	R	I	E	P
R_1	470 Ω			
R_2	2200 Ω			
R_3	10,000 Ω			
Total			20 V	

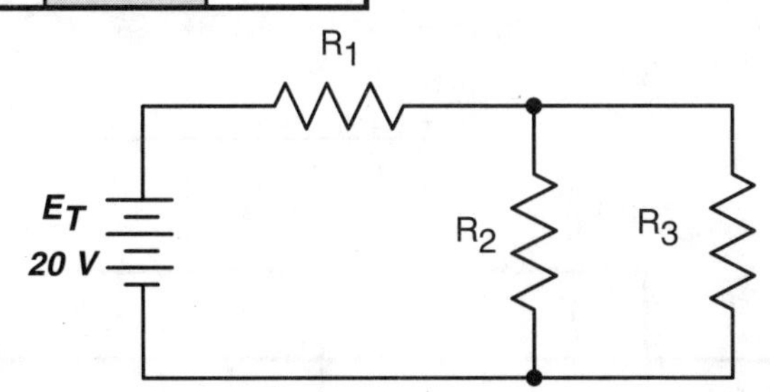

2.

	R	I	E	P
R_1	2200 Ω			
R_2	470 Ω			
R_3	1000 Ω			
Total			20 V	

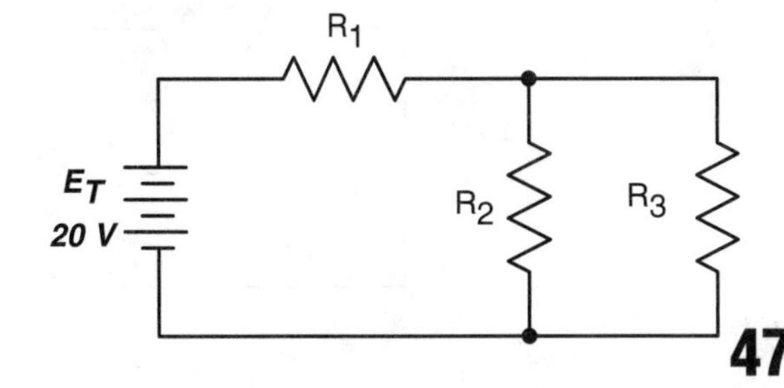

Directions

Solve for all values listed for each problem shown. Show all work and draw all equivalent circuits.

LAB 1-20

Series-Parallel DC Circuits

Ohm's Law: $I = E / R$

Formulas for series circuit:

$R_T = R_1 + R_2 + R_3 + \ldots R_n$

$E_T = E_1 + E_2 + E_3 + \ldots E_n$

$I_T = I_1 = I_2 = I_3 = \ldots I_n$

$P_T = P_1 + P_2 + P_3 + \ldots P_n$

Formulas for parallel circuit:

$\frac{1}{R_T} = \frac{1}{R_1} + \frac{1}{R_2} + \frac{1}{R_3} + \ldots \frac{1}{R_n}$

$E_T = E_1 = E_2 = E_3 = \ldots E_n$

$I_T = I_1 + I_2 + I_3 + \ldots I_n$

$P_T = P_1 + P_2 + P_3 + \ldots P_n$

3.

	R	I	E	P
R_1	470 Ω			
R_2	1000 Ω			
R_3	2200 Ω			
R_4	4700 Ω			
Total			20 V	

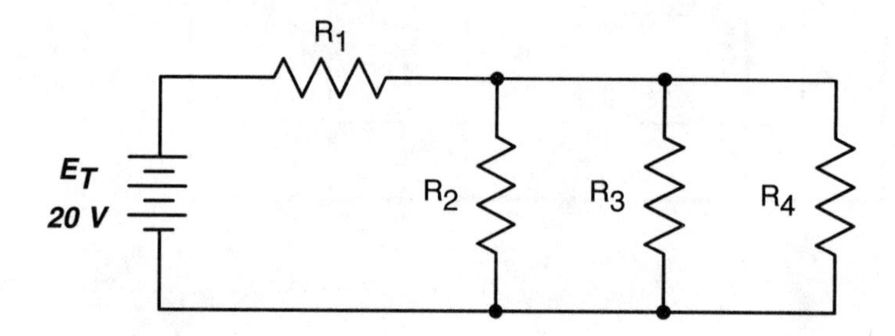

4.

	R	I	E	P
R_1	4700 Ω			
R_2	470 Ω			
R_3	1000 Ω			
R_4	2200 Ω			
Total			20 V	

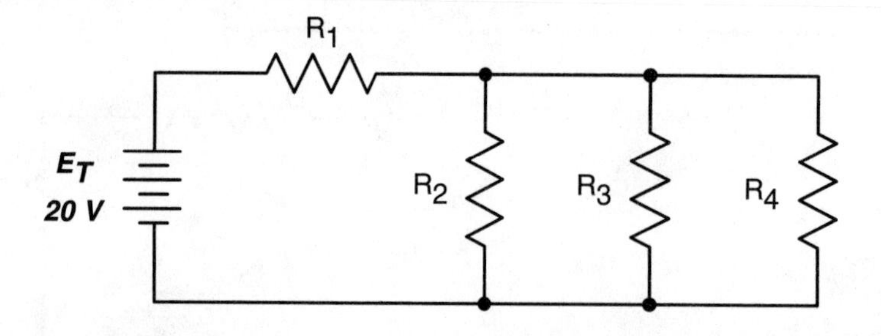

Notes

LAB 1-20

Series-Parallel DC Circuits

Name
Course
Date Due

Directions

Solve for all values listed for each problem shown. Show all work and draw all equivalent circuits.

Ohm's Law: $I = E / R$

Formulas for series circuit:

$R_T = R_1 + R_2 + R_3 + \ldots R_n$

$E_T = E_1 + E_2 + E_3 + \ldots E_n$

$I_T = I_1 = I_2 = I_3 = \ldots I_n$

$P_T = P_1 + P_2 + P_3 + \ldots P_n$

Formulas for parallel circuit:

$\frac{1}{R_T} = \frac{1}{R_1} + \frac{1}{R_2} + \frac{1}{R_3} + \ldots \frac{1}{R_n}$

$E_T = E_1 = E_2 = E_3 = \ldots E_n$

$I_T = I_1 + I_2 + I_3 + \ldots I_n$

$P_T = P_1 + P_2 + P_3 + \ldots P_n$

5.

	R	I	E	P
R_1	470 Ω			
R_2	1000 Ω			
R_3	2200 Ω			
R_4	10,000 Ω			
R_5	470 Ω			
Total			20 V	

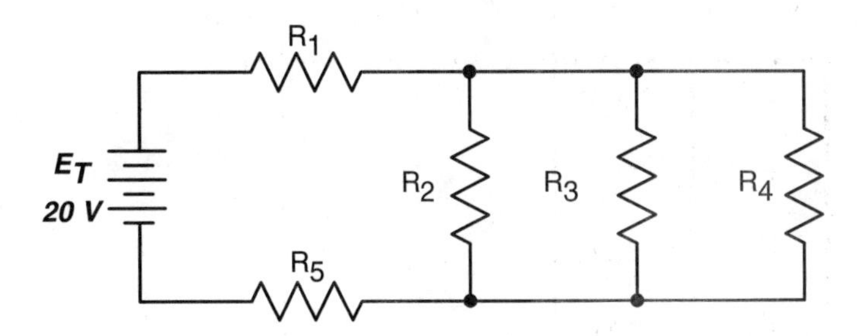

Notes

LAB 1-21

Current in a Series Circuit

Name
Course
Date Due

Objective

The student should be able to observe how current flow is the same throughout a series DC circuit.

Reference

Chapter 2, pages 10-16

Chapter 6, pages 61-73

Materials Required

Resistors

- R_1 - 470 Ω, 1/2 W, Yellow - Violet - Brown
- R_2 - 1000 Ω, 1/2 W, Brown - Black - Red
- R_3 - 2200 Ω, 1/2 W, Red - Red - Red
- R_4 - 4700 Ω, 1/2 W, Yellow - Violet - Red
- R_5 - 10,000 Ω, 1/2 W, Brown - Black - Orange

***NOTE**: The tolerance band for the resistors may be gold, silver, or none.*

Equipment Required

DC Ammeter with Leads

DC Voltmeter with Leads

Adjustable Power Supply

Notes

Safety Precautions

An ammeter is always connected in series in a circuit. Always set the ammeter on the highest range prior to applying voltage.

Procedure

As each step is completed, check it off.

1. Using Lab 1-18, transfer the calculated current through each resistor and the total current flow for each resistor combination to Table 1-21-1. Carefully check the values of R_n and each resistor combination when making this transfer.
2. Connect the circuit as shown in Figure 1-21-1 using the resistor combination from Table 1-21-1.
3. Turn the power supply on and use the voltmeter to adjust for 20 V.
4. Measure the current flow through each resistor and record it on Table 1-21-1. Remember, to measure current flow, the circuit has to be broken to insert the ammeter in it.
5. Remove power and disconnect the circuit.
6. Repeat steps 2 - 5 for each resistor combination listed on Table 1-21-1.
7. On Table 1-21-1, compare the measured current flow of the individual resistors with the calculated current flow through the individual resistors.
8. Indicate whether the calculated value is within 10 mA of the actual value measured with a Yes or No under the column marked Current Flow Comparison.

FIGURE 1-21-1

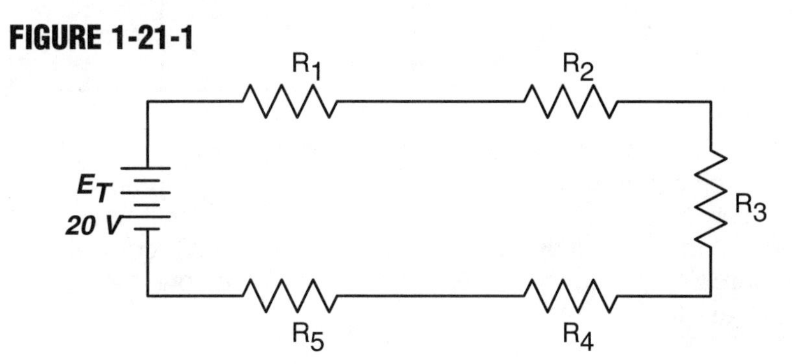

Directions

Complete Table 1-21-1 using the steps on page 51.

TABLE 1-21-1

Resistor Combination		Individual Current		Total Current		Current Flow Comparison	
		Calculated	Measured	Calculated	Measured	Indiv.	Total
1. $R_2 + R_4$	R_2						
	R_4						
2. $R_1 + R_2 + R_3$	R_1						
	R_2						
	R_3						
3. $R_2 + R_4 + R_5$	R_2						
	R_4						
	R_5						
4. $R_1 + R_3 + R_4 + R_5$	R_1						
	R_3						
	R_4						
	R_5						
5. $R_1 + R_2 + R_3 + R_4 + R_5$	R_1						
	R_2						
	R_3						
	R_4						
	R_5						

Summary

In a series circuit there is only one path for the current to flow. The total current flow in a series circuit is the same as the current flow through each of the individual resistors. Therefore, the total current flow in the circuit may be used to determine the current flow through an individual resistor in a series circuit.

LAB 1-21

Current in a Series Circuit

Notes

LAB 1-22

Voltage in a Series Circuit

Name
Course
Date Due

Objective

The student should be able to accurately measure the voltage drop across each resistor in a series DC circuit.

Reference

Chapter 3, pages 17-30

Chapter 6, pages 61-73

Materials Required

Resistors

R_1 - 470 Ω, 1/2 W, Yellow - Violet - Brown

R_2 - 1000 Ω, 1/2 W, Brown - Black - Red

R_3 - 2200 Ω, 1/2 W, Red - Red - Red

R_4 - 4700 Ω, 1/2 W, Yellow - Violet - Red

R_5 - 10,000 Ω, 1/2 W, Brown - Black - Orange

***NOTE**: The tolerance band for the resistors may be gold, silver, or none.*

Equipment Required

DC Voltmeter with Leads

Adjustable Power Supply

Notes

Safety Precautions

A voltmeter is always connected in parallel in a circuit. Always set the voltmeter on the highest range prior to applying voltage.

Procedure

As each step is completed, check it off.

1. Using Lab 1-18, transfer the calculated voltage drop for each resistor combination to Table 1-22-1. Carefully check the values of R_n and each resistor combination when making this transfer.
2. Connect the circuit as shown in Figure 1-22-1 using the resistor combination from Table 1-22-1.
3. Turn the power supply on and use the voltmeter to adjust for 20 V.
4. Measure the voltage drop across each resistor and record it on Table 1-22-1.
5. Remove power and disconnect the circuit.
6. Repeat steps 2 - 5 for each resistor combination listed on Table 1-22-1.
7. On Table 1-22-1, compare the measured voltage drop of the individual resistors with the calculated voltage drop of the individual resistors.
8. Indicate whether the calculated value is within 5 V of the actual value measured with a Yes or No under the column marked Voltage Drop Comparison.

FIGURE 1-22-1

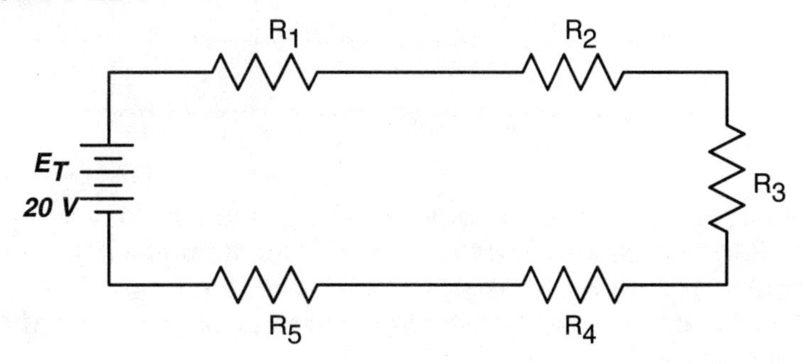

Directions

Complete Table 1-22-1 using the steps on page 53.

TABLE 1-22-1

Resistor Combination		Color Code Value	Actual Value	Voltage Drop		Voltage Drop Comparison
				Calculated	Measured	
1. $R_2 + R_4$	R_2					
	R_4					
2. $R_1 + R_2 + R_3$	R_1					
	R_2					
	R_3					
3. $R_2 + R_4 + R_5$	R_2					
	R_4					
	R_5					
4. $R_1 + R_3 + R_4 + R_5$	R_1					
	R_3					
	R_4					
	R_5					
5. $R_1 + R_2 + R_3 + R_4 + R_5$	R_1					
	R_2					
	R_3					
	R_4					
	R_5					

Summary

In a series circuit the total voltage is equal to the sum of the individual voltage drops in the circuit. In each circuit combination the voltage dropped across each resistor should have added up to the total voltage applied. In each circuit combination 20 V was applied by the power supply.

The reason there is a difference between measured voltage drop and the calculated voltage drop is the values of the resistors used. The calculated voltage drop was based on the color code values. The measured voltage drop was done using the actual resistor values.

LAB 1-22

Voltage in a Series Circuit

Notes

LAB 1-23

Current in a Parallel Circuit

Name
Course
Date Due

Objective

The student should be able to accurately measure the current flow across each of the branches of a parallel DC circuit.

Reference

Chapter 2, pages 10-16

Chapter 6, pages 61-73

Materials Required

Resistors

R_1 - 470 Ω, 1/2 W, Yellow - Violet - Brown

R_2 - 1000 Ω, 1/2 W, Brown - Black - Red

R_3 - 2200 Ω, 1/2 W, Red - Red - Red

R_4 - 4700 Ω, 1/2 W, Yellow - Violet - Red

R_5 - 10,000 Ω, 1/2 W, Brown - Black - Orange

***NOTE**: The tolerance band for the resistors may be gold, silver, or none.*

Equipment Required

DC Ammeter with Leads

DC Voltmeter with Leads

Adjustable Power Supply

Notes

Safety Precautions

An ammeter is always connected in series in a circuit. Always set the ammeter on the highest range prior to applying voltage.

Procedure

As each step is completed, check it off.

1. Using Lab 1-19, transfer the calculated current through each resistor and the total current flow for each resistor combination to Table 1-23-1. Carefully check the values of R_n and each resistor combination when making this transfer.
2. Connect the circuit as shown in Figure 1-23-1 using the resistor combination from Table 1-23-1.
3. Turn the power supply on and use the voltmeter to adjust for 20 V.
4. Measure the current flow through each resistor and record it on Table 1-23-1. Remember, to measure current flow, the circuit has to be broken to insert the ammeter in it.
5. Remove power and disconnect the circuit.
6. Repeat steps 2 - 5 for each resistor combination listed on Table 1-23-1.
7. On Table 1-23-1, compare the measured current flow of the individual resistors with the calculated current flow through the individual resistors.
8. Indicate whether the calculated value is within 10 mA of the actual value measured with a Yes or No under the column marked Current Flow Comparison.

FIGURE 1-23-1

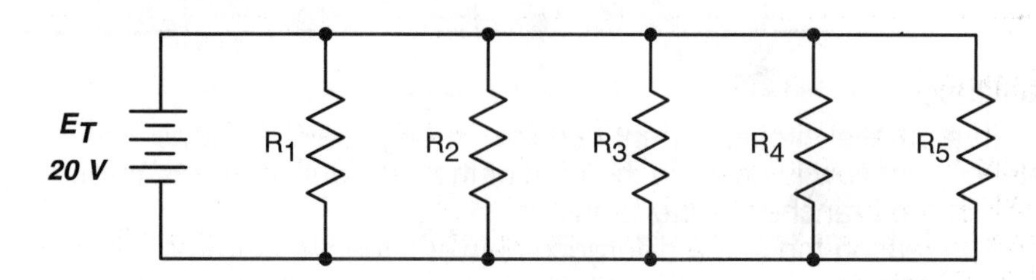

LAB 1-23

Current in a Parallel Circuit

Directions

Complete Table 1-23-1 using the steps on page 53.

TABLE 1-23-1

Resistor Combination		Individual Current		Total Current		Current Flow Comparison	
		Calculated	Measured	Calculated	Measured	Indiv.	Total
1. $R_2 \parallel R_4$	R_2						
	R_4						
2. $R_1 \parallel R_2 \parallel R_3$	R_1						
	R_2						
	R_3						
3. $R_2 \parallel R_4 \parallel R_5$	R_2						
	R_4						
	R_5						
4. $R_1 \parallel R_3 \parallel R_4 \parallel R_5$	R_1						
	R_3						
	R_4						
	R_5						
5. $R_1 \parallel R_2 \parallel R_3 \parallel R_4 \parallel R_5$	R_1						
	R_2						
	R_3						
	R_4						
	R_5						

Summary

In a parallel circuit, current can flow through several paths. The total current flow is equal to the sum of the individual currents through each of the branches of the circuit.

The reason there is a difference between the measured current and the calculated current is the value of the resistors used. The calculated current drop was based on the color code values. The measured current was accomplished using the actual resistor values.

Notes

LAB 1-24

Voltage in a Parallel Circuit

Name

Course

Date Due

Objective

The student should be able to accurately measure the voltage drop across each of the branches in a parallel DC circuit.

Reference

Chapter 3, pages 17-30

Chapter 6, pages 61-73

Materials Required

Resistors

R_1 - 470 Ω, 1/2 W, Yellow - Violet - Brown

R_2 - 1000 Ω, 1/2 W, Brown - Black - Red

R_3 - 2200 Ω, 1/2 W, Red - Red - Red

R_4 - 4700 Ω, 1/2 W, Yellow - Violet - Red

R_5 - 10,000 Ω, 1/2 W, Brown - Black - Orange

***NOTE**: The tolerance band for the resistors may be gold, silver, or none.*

Equipment Required

DC Voltmeter with Leads

Adjustable Power Supply

Notes

Safety Precautions

A voltmeter is always connected in parallel in a circuit. Always set the voltmeter on the highest range prior to applying voltage.

Procedure

As each step is completed, check it off.

1. Using Lab 1-19, transfer the calculated voltage drop for each resistor combination to Table 1-24-1. Carefully check the values of R_n and each resistor combination when making this transfer.
2. Connect the circuit as shown in Figure 1-24-1 using the resistor combination from Table 1-24-1.
3. Turn the power supply on and use the voltmeter to adjust for 20 V.
4. Measure the voltage drop across each resistor and record it on Table 1-24-1.
5. Remove power and disconnect the circuit.
6. Repeat steps 2 - 5 for each resistor combination listed on Table 1-24-1.
7. On Table 1-24-1, compare the measured voltage drop of the individual resistors with the calculated voltage drop of the individual resistors.
8. Indicate whether the calculated value is within 5 V of the actual value measured with a Yes or No under the column marked Voltage Drop Comparison.

FIGURE 1-24-1

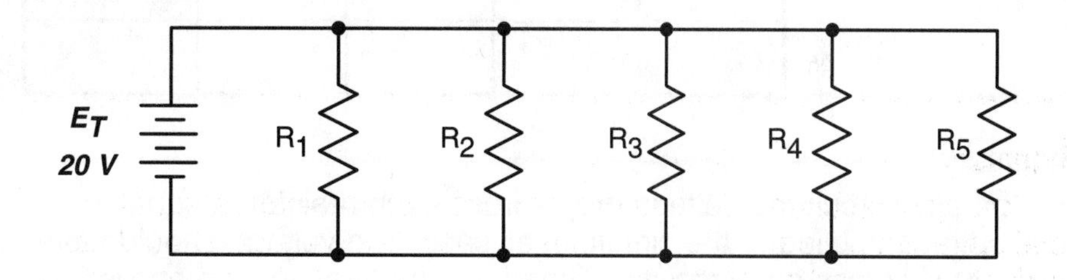

Directions

Complete Table 1-24-1 using the steps on page 57.

TABLE 1-24-1

Resistor Combination		Color Code Value	Actual Value	Voltage Drop		Voltage Drop Comparison
				Calculated	Measured	
1. $R_2 \parallel R_4$	R_2					
	R_4					
2. $R_1 \parallel R_2 \parallel R_3$	R_1					
	R_2					
	R_3					
3. $R_2 \parallel R_4 \parallel R_5$	R_2					
	R_4					
	R_5					
4. $R_1 \parallel R_3 \parallel R_4 \parallel R_5$	R_1					
	R_3					
	R_4					
	R_5					
5. $R_1 \parallel R_2 \parallel R_3 \parallel R_4 \parallel R_5$	R_1					
	R_2					
	R_3					
	R_4					
	R_5					

Summary

In a parallel circuit the voltage drop across each resistor is equal to the total voltage applied to the circuit. In all cases the voltage should have been 20 V across each resistor. Therefore, the total voltage applied to the circuit may be used to determine the voltage drop across each individual resistor in a parallel DC circuit.

LAB 1-24

Voltage in a Parallel Circuit

Notes

LAB 1-25

Magnetism

Name
Course
Date Due

Objective

The student should be able to identify the terminology associated with magnetism.

Reference

Chapter 9, pages 94-105

Materials Required

None

Equipment Required

None

Notes

Definitions

Define the following terms in complete sentences.

1. Magnet
2. Artificial Magnet
3. Electromagnet
4. Magnetism
5. Domains
6. Flux Lines
7. Permeability
8. Magnetic Induction
9. Retentivity
10. Faraday's Law
11. Relay
12. Solenoid

Questions

Answer the following questions in complete sentences.

1. What is the domain theory?

2. What two factors affect the strength of an electromagnet?

LAB 1-25

Magnetism

Notes

LAB 1-26

Inductance

Name

Course

Date Due

Objective

The student should be able to identify the terminology associated with inductance.

Reference

Chapter 10, pages 106-111

Materials Required

None

Equipment Required

None

Notes

Definitions

Define the following terms in complete sentences.

1. Inductance

2. Henry

3. Inductor

4. Choke

5. Domains

6. Time Constant

Questions

Answer the following questions in complete sentences.

1. What letter symbol is used to represent inductance?

2. What is the unit of inductance?

3. What letter symbol represents the unit of inductance?

4. Draw the schematic symbol for an air-core inductor.

5. Draw the schematic symbol for an iron-core inductor.

LAB 1-26

Inductance

Notes

LAB 1-27

Capacitance

Objective

The student should be able to identify the terminology associated with capacitance.

Reference

Chapter 11, pages 112-118

Materials Required

None

Equipment Required

None

Notes

Name

Course

Date Due

Definitions

Define the following terms in complete sentences.

1. Capacitance
2. Plate
3. Dielectric
4. Farad
5. Microfarad
6. Picofarad
7. Electrolytic Capacitor
8. Paper Capacitor
9. Ceramic Disk Capacitor
10. Padder Capacitor
11. Trimmer Capacitor

Questions

Answer the following questions in complete sentences.

1. What letter symbol represents capacitance?

2. What is the unit of capacitance?

3. What letter symbol represents the unit of capacitance?

4. Draw the schematic symbol for an electrolytic capacitor.

5. Describe the process for determining an RC time constant.

LAB 1-27

Capacitance

Notes

LAB 1-28

Capacitor Application

Objective

The student should be able to demonstrate how a capacitor reacts in a circuit.

Reference

Chapter 11, pages 112-118

Materials Required

470 μF, 50 V Electrolytic Capacitor
6 V Light Bulb and Socket
Hookup Wire

Equipment Required

Adjustable Power Supply
DC Analog Voltmeter with Leads

Notes

Name
Course
Date Due

Procedure

As each step is completed, check it off.

1. Connect the circuit as shown in Figure 1-28-1.
2. Adjust the power supply for 6 V.
3. Move switch S_1 back and forth several times. Does the bulb light? If it does, does it stay lit?
4. Move switch S_1 to the A position. Does any current flow through the light after the first instant?
5. Remove power and disconnect the circuit.
6. Connect the circuit as shown in Figure 1-28-2.
7. Adjust the power supply for 10 V.
8. Open and close switch S_1 at a rate of once per second. Notice the meter fluctuates between 0 and 10 V.
9. Add the capacitor as shown in Figure 1-28-3.
10. Again, open and close switch S_1 at a rate of once per second. Does the meter behave differently?
11. Remove power and disconnect the circuit.

Procedure

Use these schematics for the steps on page 65.

FIGURE 1-28-1

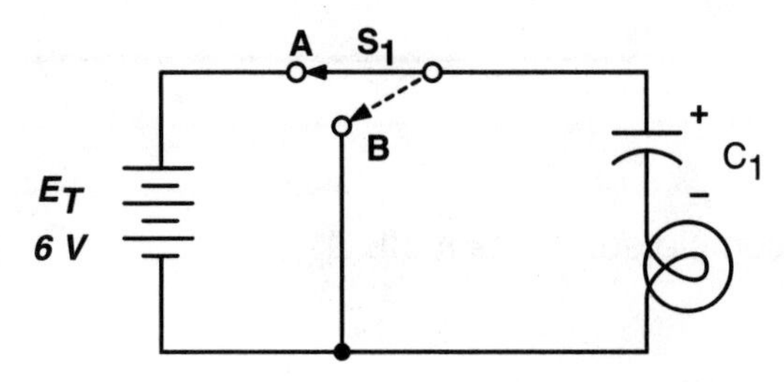

FIGURE 1-28-2

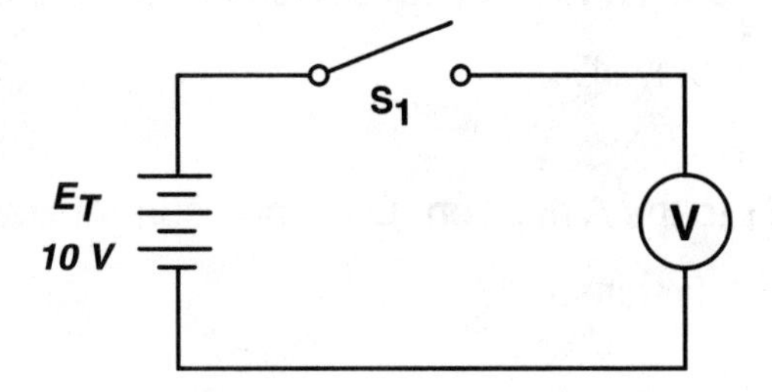

FIGURE 1-28-3

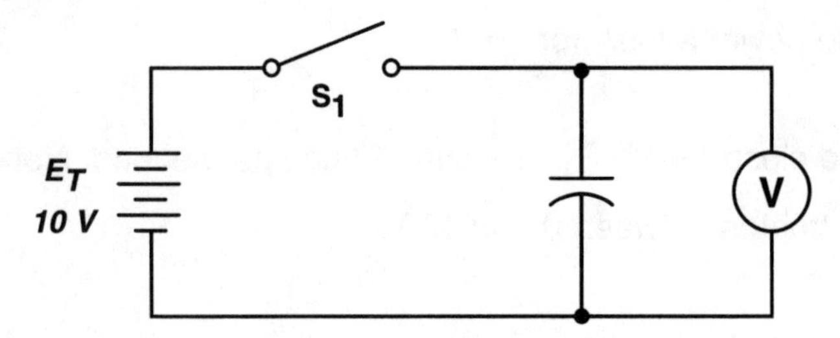

Summary

The light bulb lit momentarily each time the switch was moved back and forth. This occurred because when the switch was in the A position, the capacitor charged up to the power supply voltage. Initially the light lit for a brief instant as current was flowing to charge the capacitor. When the switch was moved to the B position, the capacitor discharged quickly through the light. This resulted in the light bulb lighting again for an instant.

Steps 6 - 10 demonstrate the purpose of a capacitor. The capacitor converted the pulsating DC signal to a steady DC signal. When the switch was closed, the capacitor charged up. When the switch was opened, the capacitor discharged through the load, in this case the meter. This technique is used in power supplies to provide a steady DC output.

LAB 1-28

Capacitor Application

Notes

LAB 2-1

Alternating Current

Name

Course

Date Due

Objective

The student should be able to identify the terminology associated with alternating current.

Reference

Chapter 12, pages 121-128

Materials Required

None

Equipment Required

None

Notes

Definitions

Define the following terms in complete sentences.

1. Alternation
2. Cycle
3. Hertz
4. Sine Wave
5. Sinusoidal Waveform
6. Amplitude
7. Peak Value
8. Peak-to-Peak Value
9. Effective Value
10. RMS Value
11. Period
12. Frequency

Definitions

Define the following terms in complete sentences.

13. Nonsinusoidal Waveforms

14. Pulse Width

15. Fundamental Frequency

16. Harmonics

Questions

Answer the following questions in complete sentences.

1. What letter symbol is used to represent alternating current?

2. Draw a sine wave.

3. Draw three different nonsinusoidal waves.

LAB 2-1

Alternating Current

Notes

LAB 2-2

AC Measurements

Name

Course

Date Due

Objective

The student should be able to identify the terminology associated with alternating current.

Reference

Chapter 13, pages 129-137

Materials Required

None

Equipment Required

None

Notes

Questions

Answer the following questions in complete sentences.

1. What are three types of AC meters?

   a.

   b.

   c.

2. Identify the basic blocks of an oscilloscope.

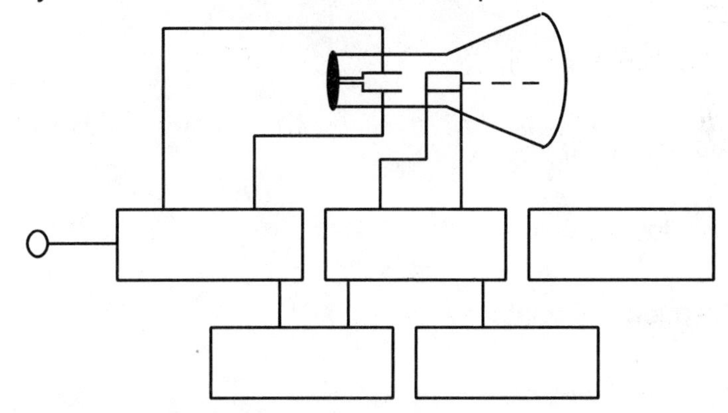

3. List the steps for initially setting up an oscilloscope.

   Intensity

   Focus

   Astigmatism

   Vertical Position

   Horizontal Position

   Triggering

   Level

   Time/cm

   Volt/cm

Questions

Answer the following questions in complete sentences.

4. Draw and identify the basic blocks of a frequency counter.

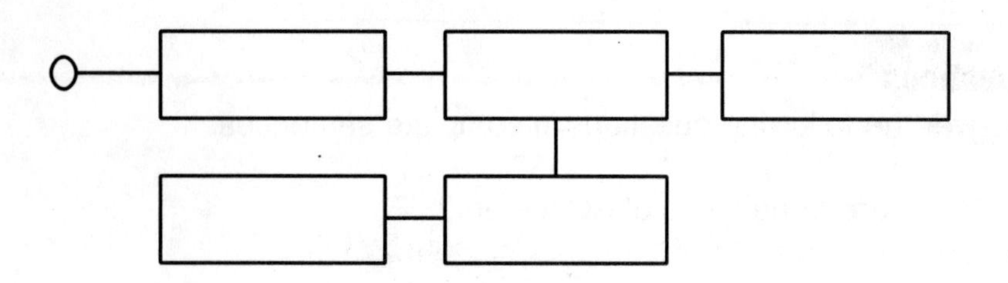

5. Identify the function of each of the following pieces of test equipment.

   AC Ammeter

   AC Voltmeter

   Oscilloscope

   Frequency Counter

Notes

LAB 2-3

AC Measurements: RMS, Peak, and Peak-to-Peak Voltage

Name
Course
Date Due

Objective

The student should be able to describe and accurately measure RMS, peak, and peak-to-peak voltages.

Reference

Chapter 13, pages 129-137

Materials Required

Resistors

R_1 - 470 Ω, 1/2 W, Yellow-Violet-Brown

R_2 - 1000 Ω, 1/2 W, Brown-Black-Red

Equipment Required

Low Voltage AC Power Supply

AC Voltmeter

Notes

Procedure

As each step is completed, check it off.

1. Connect the circuit as shown in Figure 2-3-1.
2. Connect the AC voltmeter to measure the total voltage across resistors R_1 and R_2.
3. Record the total AC voltage on Table 2-3-1.
4. Calculate the peak value of the AC voltage read on the voltmeter using the following formula:

$$E_{RMS} = E_P \times 0.707$$

5. Record the results on Table 2-3-1.
6. Calculate the peak-to-peak value of the AC reading using the following formula:

$$E_{P-P} = E_P \times 2$$

7. Record the results on Table 2-3-1.
8. Disconnect the AC voltmeter and reconnect it to measure the voltage drop across resistor R_1.
9. Record the AC voltage on Table 2-3-1.
10. Calculate the peak value of the voltage drop across resistor R_1 and record it on Table 2-3-1.
11. Calculate the peak-to-peak value of the voltage drop across resistor R_1 and record it on Table 2-3-1.

Procedure

As each step is completed, check it off.

12. Disconnect the AC voltmeter from across resistor R_1 and reconnect it to measure the voltage drop across resistor R_2.
13. Record the AC voltage on Table 2-3-1.
14. Calculate the peak value of the voltage drop across resistor R_2 and record it on Table 2-3-1.
15. Calculate the peak-to-peak value of the voltage drop across resistor R_2 and record it on Table 2-3-1.

Figure 2-3-1

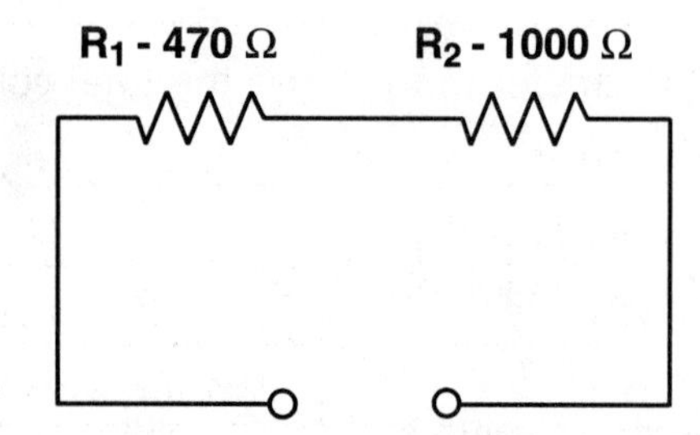

Table 2-3-1

	E_{RMS}	E_P	$E_{P\text{-}P}$
R_1			
R_2			
Total			

Summary

The RMS value of an AC voltage is the value always referred to unless otherwise specified. The RMS value is the same as the effective value. The RMS value will cause the same heating effect as a DC voltage of the same value. All AC voltmeters have RMS scales. Some AC voltmeters may also have a peak-to-peak scale. The peak-to-peak voltage is the value from the positive peak to the negative peak or two times the peak value. The peak value of an AC voltage is the value of the positive or negative alternation.

Notes

LAB 2-4

AC Measurements: Oscilloscope

Name
Course
Date Due

Objective

The student should be able to set up an oscilloscope and accurately measure an AC signal.

Reference

Chapter 13, pages 129-137

Materials Required

None

Equipment Required

Low Voltage AC Power Supply with Leads

Oscilloscope

Notes

Procedure

As each step is completed, check it off.

1. Initially set the controls on the oscilloscope as follows:

Trigger Source	Channel 1
Trigger Slope	+
Trigger Coupling	AC
Time/cm	0.5 ms/calibrated
Volt/cm	50 mV/calibrated/Pushed In
AC/DC	AC
Vertical Mode	Channel 1
Intensity	Mid-Position
Focus	Mid-Position
Vertical Position	Mid-Position
Horizontal Position	Mid-Position/Pushed In

NOTE*: The controls may have a different label based on the type of oscilloscope used. Check the manufacturer's operation manual if there is a question.*

2. Connect the oscilloscope probe to the channel 1 (dual-trace oscilloscope) vertical input jack.

3. Connect the probe to the "calibrate probe" test point. Make sure the probe is in the x10 position.

4. Adjust the controls until a stable full-screen display of two to three cycles is obtained.

5. Connect the scope probe and ground to the leads of the low voltage AC power supply.

6. Turn on the power supply. Set to 12 V AC.

7. Adjust the controls to two to three cycles and draw an accurate representation of the wave form on Figure 2-4-1.

8. Determine the peak-to-peak value of the input signal and the time required to complete one cycle and record the results on Table 2-4-1.

Procedure

As each step is completed, check it off.

9. Convert the time for one cycle to frequency in hertz (cycle per seconds) using the formula:

$$f = \frac{1}{t}$$

10. Using the peak-to-peak value, determine the peak value of the signal using the formula:

$$E_P = \frac{E_{P-P}}{2}$$

11. Record the results on Table 2-4-1.

12. Using the results from step 10, determine the RMS value of the signal using the formula:

$$E_{RMS} = E_P \times 0.707$$

13. Record the results on Table 2-4-1.

Summary

The oscilloscope can be used to measure important AC values and observe various types of AC waveforms.

To measure the amplitude of a signal, the volts-per-centimeter control is adjusted for a setting that provides full signal viewing. The number of centimeters the signal falls across is counted and multiplied by the volt/cm setting. This value represents the peak-to-peak value of the signal.

The oscilloscope can also be used to determine the period of an AC waveform. The period, time for one cycle, is determined by observing the horizontal width of the waveform on the screen. The sweep generator, time/cm control, is adjusted so the electron beam moves across the screen at a specific rate. The time/cm control sets the amount of time in seconds, milliseconds, or microseconds. Each centimeter of horizontal deflection represents a time interval determined by the time/cm control.

The frequency of an AC waveform can be determined by measuring the period of the waveform and then calculating the frequency.

Notes

LAB 2-4

AC Measurements: Oscilloscope

Name	
Course	
Date Due	

Notes

Directions

Complete Figure 2-4-1 and Table 2-4-1 using the steps on pages 73 and 74.

FIGURE 2-4-1

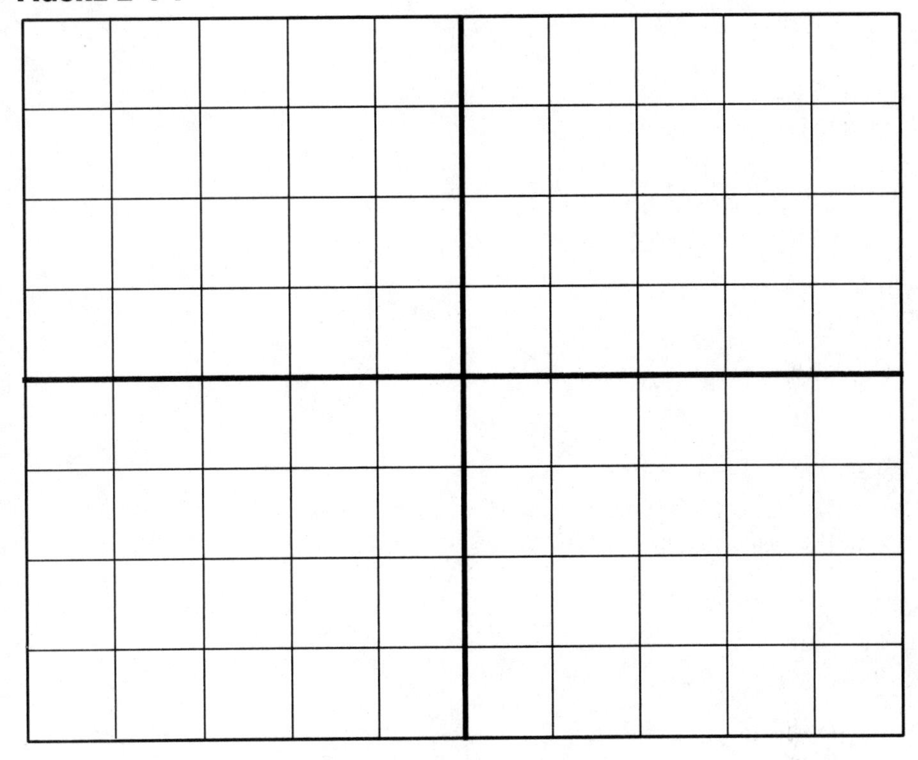

TABLE 2-4-1

t	f	E_{RMS}	E_P	E_{P-P}

LAB 2-4

AC Measurements: Oscilloscope

Notes

LAB 2-5

Capacitive AC Circuits

Name

Course

Date Due

Objective

The student should be able to explain the effects of capacitors in AC circuits.

Reference

Chapter 15, pages 145-151

Materials Required

None

Equipment Required

None

Notes

Directions

Draw the required circuits and diagrams.

1. Draw a sine wave diagram showing the relationship between current and voltage in a capacitive AC circuit.

2. Draw a low-pass filter circuit using capacitors and resistors.

3. Draw a high-pass filter circuit using capacitors and resistors.

Directions

Draw the required circuits and diagrams.

4. Draw a circuit that would be used to allow a DC signal to pass but would eliminate an AC signal.

5. Draw a circuit that would pass an AC signal but would block a DC signal.

LAB 2-5

Capacitive AC Circuits

Notes

LAB 2-6

Capacitors in AC Circuits

Objective

The student should be able to solve for capacitive reactance in a capacitive AC circuit.

Reference

Chapter 16, pages 145-151

Materials Required

None

Equipment Required

Calculator

Notes

Name

Course

Date Due

Directions

Solve for capacitive reactance in each of the following problems. Show all work.

Formula for capacitive reactance:

$$X_C = \frac{1}{2\pi fC}$$

Where:

X_C = capacitive reactance in ohms
π = pi, the constant 3.14
f = frequency in hertz
C = capacitance in farads

1. What is the capacitive reactance of a 1000 µF capacitor at 60 Hz?

2. What is the capacitive reactance of a 1000 µF capacitor at 120 Hz?

LAB 2-6

Capacitors in AC Circuits

Notes

LAB 2-7

RC Circuits

Objective

The student should be able to demonstrate the characteristics of a series RC network.

Reference

Chapter 15, pages 145-151

Materials Required

4700 Ω, 1/2 W Resistor, Yellow- Violet - Red
0.47 μF Capacitor
47 μF Capacitor
Sheet of Semilogarithmic Graph Paper (3 cycles x 10 to the inch)

Equipment Required

Signal or Function Generator
AC Voltmeter

Notes

Name
Course
Date Due

Procedure

As each step is completed, check it off.

1. Connect the circuit shown in Figure 2-7-1 using the 0.47 μF capacitor.
2. Connect the AC voltmeter across the capacitor.
3. Turn on the signal generator and set the amplitude knob to the mid-position of its range.
4. Set the frequency dial of the signal generator to the first position indicated on Table 2-7-1.
5. Record the voltage read on the AC voltmeter on Table 2-7-1.
6. Set the frequency dial on the signal generator to the next position and record the voltage reading on Table 2-7-1.
7. Repeat step 6 for the rest of the frequencies on Table 2-7-1.
8. Turn off the signal generator.
9. Replace the 0.47 μF capacitor with the 47 μF capacitor.
10. Repeat steps 1 - 8 for the 47 μF capacitor.
11. Plot the results using the semilogarithmic graph paper.
12. Compare the results of the two capacitors and write a conclusion.

LAB 2-7

RC Circuits

Notes

Directions

Complete Table 2-7-1.

FIGURE 2-7-1

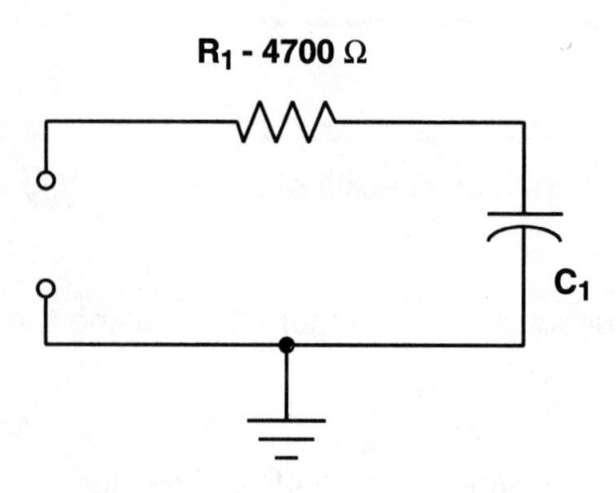

TABLE 2-7-1

Frequency	E_{C1}	E_{C2}
500 Hz		
1000 Hz		
2500 Hz		
5000 Hz		
10,000 Hz		

Summary

This demonstrates the effect of a changing frequency on a circuit reactance. Increasing the frequency caused a decrease in the voltage across the capacitor. The voltage across the capacitor is proportional to the value of the capacitive reactance. The decrease in voltage represented a decrease in reactance. This demonstrates that reactance is inversely proportional to frequency.

By increasing the size of the capacitor in the circuit, the voltage decreased.
Again, the voltage across the capacitor is proportional to the value of the capacitive reactance. This demonstrates that the reactance is inversely proportional to the size of the capacitor.

LAB 2-8

Inductive AC Circuits

Name

Course

Date Due

Objective

The student should be able to explain the effects of inductors in AC circuits.

Reference

Chapter 16, pages 152-157

Materials Required

None

Equipment Required

None

Notes

Directions

Draw the required circuits and diagrams.

1. Draw a sine wave diagram showing the relationship between current and voltage in an inductive AC circuit.

2. Draw a low-pass filter circuit using inductors and resistors.

3. Draw a high-pass filter circuit using inductors and resistors.

LAB 2-8

Inductive AC Circuits

Notes

LAB 2-9

Inductors in AC Circuits

Name

Course

Date Due

Objective

The student should be able to solve for inductive reactance in an inductive AC circuit.

Reference

Chapter 16, pages 152-157

Materials Required

None

Equipment Required

None

Notes

Directions

Solve for inductive reactance in each of the following problems. Show all work.

Formula for inductive reactance:

$X_L = 2\pi fL$

Where:

X_L = inductive reactance in ohms
π = pi, the constant 3.14
f = frequency in hertz
L = inductance in henries

1. What is the inductive reactance of a 10 mH inductor at 60 Hz?

2. What is the inductive reactance of a 10 mH inductor at 120 Hz?

LAB 2-9

Inductors in AC Circuits

Notes

LAB 2-10

RL Circuits

Objective

The student should be able to demonstrate the characteristics of a series RL network.

Reference

Chapter 16, pages 152-157

Materials Required

4700 Ω, 1/2 W Resistor, Yellow- Violet- Red

10 mH Iron-Core Choke

Sheet of Semilogarithmic Graph Paper (3 cycles x 10 to the inch)

Equipment Required

Signal or Function Generator

AC Voltmeter

Notes

Name
Course
Date Due

Procedure

As each step is completed, check it off.

1. Connect the circuit shown in Figure 2-10-1.
2. Connect the AC voltmeter across the inductor.
3. Turn on the signal generator and set the amplitude knob to the mid-position of its range.
4. Set the frequency dial of the signal generator to the first position indicated on Table 2-10-1.
5. Record the voltage read on the AC voltmeter on Table 2-10-1.
6. Set the frequency dial on the signal generator to the next position and record the voltage reading on Table 2-10-1.
7. Repeat step 6 for the rest of the frequencies on Table 2-10-1.
8. Turn off the signal generator.
9. Plot the results using the semilogarithmic graph paper.

Directions

Complete Table 2-10-1.

FIGURE 2-10-1

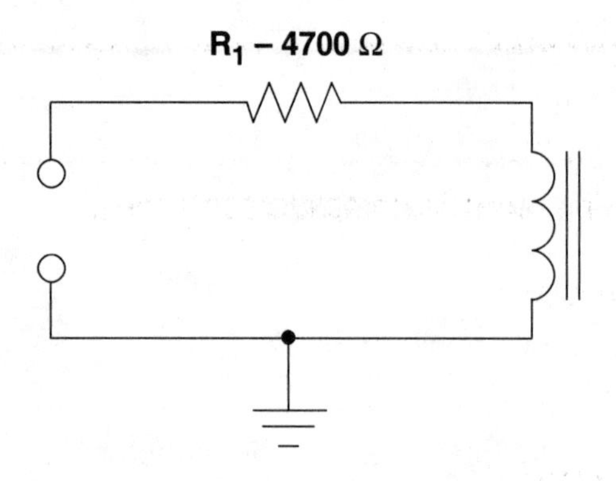

TABLE 2-10-1

Frequency	E_L
500 Hz	
1000 Hz	
2500 Hz	
5000 Hz	
10,000 Hz	

Summary

This demonstrates the effect of a changing frequency on a circuit reactance. Increasing the frequency caused an increase in the voltage across the inductor. The voltage across the inductor is proportional to the value of the inductive reactance. The increase in voltage represented an increase in reactance. This demonstrates that inductive reactance is directly proportional to the frequency.

LAB 2-10

RL Circuits

Notes

LAB 2-11

Resonance Circuits

Name

Course

Date Due

Objective

The student should be able to identify the terminology associated with resonance circuits.

Reference

Chapter 17, pages 158-166

Materials Required

None

Equipment Required

None

Notes

Definitions

Define the following terms in complete sentences.

1. Capacitor Reactance

2. Inductor Reactance

3. Impedance

Questions

Answer the following questions in complete sentences.

1. Write the formula for capacitive reactance.

2. Write the formula for inductive reactance.

3. Write the formula for reactance.

Notes

LAB 2-12

Reactance

Name

Course

Date Due

Objective

The student should be able to solve circuit reactance involving LC circuits.

Reference

Chapter 17, pages 158-166

Materials Required

None

Equipment Required

Calculator

Notes

Directions

Solve for capacitive reactance in each of the following problems. Show all work.

Formula for capacitive reactance:

$$X_C = \frac{1}{2\pi fC}$$

Formula for inductive reactance:

$$X_L = 2\pi fL$$

Formula for reactance:

$$X = X_C - X_L$$

or

$$X = X_L - X_C$$

1. What is the reactance of a circuit containing a capacitor of 1000 μF in series with an inductor of 10 mH operating at 60 Hz?

2. What is the reactance of a circuit containing a capacitor of 1 μF in series with an inductor of 250 πH operating at 10,000 Hz?

LAB 2-12

Reactance

Notes

LAB 2-13

RLC Circuits

Name

Course

Date Due

Objective

The student should be able to solve RLC circuits for impedance.

Reference

Chapter 17, pages 158-166

Materials Required

None

Equipment Required

Calculator

Notes

Directions

Solve for capacitive reactance in each of the following problems. Show all work.

Formula for capacitive reactance:

$$X_C = \frac{1}{2\pi fC}$$

Formula for inductive reactance:

$$X_L = 2\pi fL$$

Formula for reactance:

$$X = X_C - X_L$$

or

$$X = X_L - X_C$$

Formula for impedance:

$$Z^2 = R^2 + X^2$$

1. What is the impedance of a circuit containing a capacitor of 470 μF in series with an inductor of 270 πH and a resistor of 1 Ω at 0.85 Hz?

2. What is the impedance of a circuit containing a capacitor of 33 μF in series with an inductor of 0.00005 πH and a resistor of 0.047 Ω at 10,000 Hz?

LAB 2-13

RLC Circuits

Notes

LAB 2-14

Transformers

Name

Course

Date Due

Objective

The student should be able to identify the terminology associated with transformers.

Reference

Chapter 18, pages 167-174

Materials Required

None

Equipment Required

None

Notes

Definitions

Define the following terms in complete sentences.

1. Electromagnetic Induction
2. Transformer
3. Primary Winding
4. Secondary Winding
5. Coefficient of Coupling
6. Volt-Amperes
7. Mutual Inductance
8. Turns Ratio
9. Step-up Transformer
10. Step-down Transformer
11. Isolation Transformer
12. Autotransformer

Questions

Answer the following questions in complete sentences.

1. Draw the symbol for an iron-core transformer.

2. What are seven uses for a transformer?

   1.

   2.

   3.

   4.

   5.

   6.

   7.

LAB 2-14

Transformers

Notes

LAB 2-15

Transformer Ratios

Name

Course

Date Due

Objective

The student should be able to identify characteristics of transformers.

Reference

Chapter 18, pages 167-174

Materials Required

2, 1k Ω, 1/2 W Resistor, Brown - Black- Red

Center-tapped, Audio Transformer

Equipment Required

AC Power Supply

Ohmmeter with Leads

AC Voltmeter

Notes

Safety Precautions

Do not touch the leads from the power supply to the transformer when turning off the power supply.

Procedure

As each step is completed, check it off.

1. Using Figure 2-15-1, identify the five leads of the transformer and label each of them with tape.
2. Using the ohmmeter, measure the DC resistance of each set of windings and record the result on Table 2-15-1.
3. Connects leads 1 and 2 to the AC power supply and connect the voltmeter to leads 1 and 2.
4. Apply power to the AC power supply and record the voltmeter reading on Table 2-15-2.
5. Turn off the power supply and reconnect the voltmeter to pins 3 and 5.
6. Turn on the power supply and record the voltmeter reading on Table 2-15-2.
7. Turn off the power supply.
8. Using the data obtained in steps 3 - 7, determine if the transformer is a step-up or step-down transformer.
9. Determine the primary-to-secondary turns ratio of the transformer using the following formula:

$$\frac{E_P}{E_S} = \frac{N_P}{N_S}$$

10. Connect leads 3 and 5 to the AC power supply and connect the voltmeter to leads 3 and 5.
11. Apply power to the AC power supply and record the voltmeter readings on Table 2-15-2.

Safety Precautions

Do not touch the leads from the power supply to the transformer when turning off the power supply.

Procedure

As each step is completed, check it off.

12. Turn off the power supply and reconnect the voltmeter to pins 1 and 2.
13. Turn on the power supply and record the voltmeter reading on Table 2-15-2.
14. Turn off the power supply.
15. Using the data obtained in steps 10 - 14, determine if the transformer is a step-up or step-down transformer.
16. Determine the primary-to-secondary turns ratio of the transformer using the following formula:

$$\frac{E_P}{E_S} = \frac{N_P}{N_S}$$

17. Connect leads 4 and 5 to the AC power supply and the voltmeter to leads 4 and 5.
18. Apply power to the AC power supply and record the voltmeter reading on Table 2-15-2.
19. Turn off the power supply and reconnect the voltmeter to leads 1 and 2.
20. Turn on the power supply and record the voltmeter reading on Table 2-15-2.
21. Turn off the power supply.
22. Using the data obtained in steps 17 - 21, determine if the transformer is a step-up or step-down transformer.
23. Determine the primary-to-secondary turns ratio of the transformer using the following formula:

$$\frac{E_P}{E_S} = \frac{N_P}{N_S}$$

Notes

LAB 2-15

Transformer Ratios

Name

Course

Date Due

Notes

Directions

Use Figure 2-15-1 for step 1 and complete Tables 2-15-1 and 2-15-2 using the steps on pages 97 and 98.

FIGURE 2-15-1

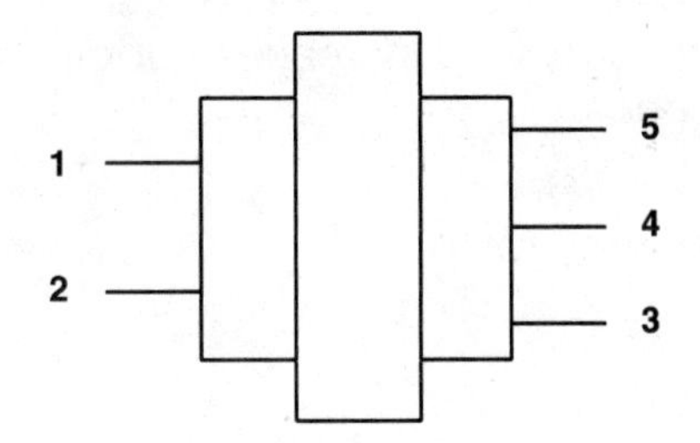

TABLE 2-15-1

Terminals	Resistance
1,2	
3,4	
4,5	
3,5	

TABLE 2-15-2

Terminals	Voltage
1,2	
3,5	
3,5	
1,2	
4,5	
1,2	

Summary

The resistance between leads 3 and 4 was equal to the resistance between leads 4 and 5. Therefore, it can be assumed that lead 4 is located at the center of the windings. The measurements taken were DC resistance measurements. In reality the AC resistance will be much higher. The AC resistance will change with frequency.

The transformer was initially hooked up as a step-down transformer. Because the voltage is stepped-down, the primary has more turns than the secondary. The ratio works out to be approximately 2.5 to 1. The transformer was then reversed and connected as a step-up transformer. The secondary had approximately 2.5 times as many turns as the primary.

In steps 19 - 23, only half of the windings were used. This resulted in the turns ratio being doubled from the previous steps. Up to this point, the losses in the transformer did not interfere with the measurements. In normal usage, a transformer will have a load on the secondary. This secondary load increases current flow in both the primary and the secondary.

Notes

LAB 3-1

Semiconductor Fundamentals

Name

Course

Date Due

Objective

The student should be able to identify the terminology associated with semiconductors.

Reference

Chapter 19, pages 177-183

Materials Required

None

Equipment Required

None

Notes

Definitions

Define the following terms in complete sentences.

1. Germanium
2. Silicon
3. Intrinsic Material
4. Covalent Bonding
5. Negative Temperature Coefficient
6. Hole
7. Electron-hole Pair
8. Doping
9. Pentavalent
10. Trivalent
11. Donor Atom
12. Majority Carrier

Definitions

Define the following terms in complete sentences.

13. Minority Carrier

14. N-Type Material

15. P-Type Material

Questions

Answer the following questions in complete sentences.

1. Describe the process for doping N-type material.

2. What makes P-type material different from N-type material?

LAB 3-1

Semiconductor Fundamentals

Notes

LAB 3-2

PN Junction Diodes

Name

Course

Date Due

Objective

The student should be able to identify the terminology and characteristics associated with PN junction diodes.

Reference

Chapter 20, pages 184-191

Materials Required

None

Equipment Required

None

Notes

Definitions

Define the following terms in complete sentences.

1. Diode
2. Depletion Region
3. Barrier Voltage
4. Bias Voltage
5. Forward Bias
6. Reverse Bias
7. Peak Inverse Voltage
8. Cathode
9. Anode

Directions

Draw the required circuits and diagrams.

1. Draw the schematic symbol for a PN junction diode and label all parts.

2. Draw a diode with a battery and resistor connected in forward bias (use schematic symbols).

3. Draw a diode with a battery and resistor connected in reverse bias (use schematic symbols).

LAB 3-2

PN Junction Diodes

Notes

LAB 3-3

Testing PN Junction Diodes

Objective
The student should be able to accurately test PN junction diodes.

Reference
Chapter 20, pages 189-190

Materials Required
5 Assorted PN junction diodes

Equipment Required
Analog Ohmmeter
or
Digital Ohmmeter with Diode Testing Capabilities

Notes

Name

Course

Date Due

Safety Precautions
Check the operating manual of the ohmmeter to ensure that it does not produce a high battery voltage at the terminals. A high battery voltage could destroy the PN junction of the diode.

Procedure
As each step is completed, check it off.

1. Identify the leads with each diode package, cathode and anode.
2. Record the identification number of each diode on Table 3-3-1.
3. Select the first diode listed on Table 3-3-1.
4. Select the ohmmeter range as follows:

   Digital Ohmmeter - 20 kΩ unless meter has a special diode test position

   Analog Ohmmeter - R x 100
5. Connect the positive or red lead of the ohmmeter to the cathode of the diode and the negative or black lead of the ohmmeter to the anode of the diode.
6. Record the resistance reading on Table 3-3-1.
7. Reverse the leads to the diode, positive to anode and negative to cathode, and record the results on Table 3-3-1.
8. Repeat steps 4 - 7 for the remaining diodes listed on Table 3-3-1.

FIGURE 3-3-1

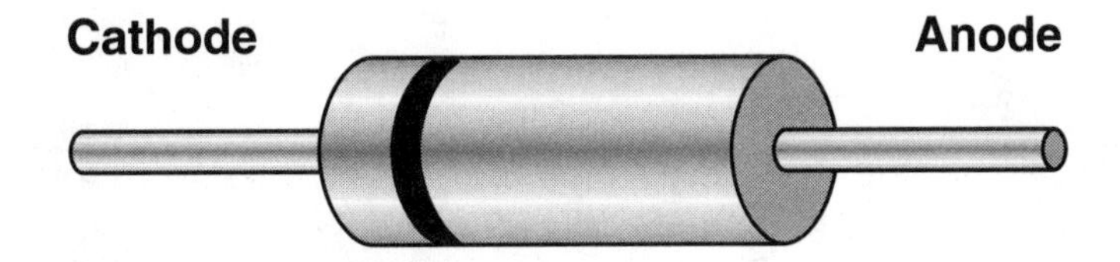

Directions

Complete Table 3-3-1 using steps 4 - 7 on page 105.

TABLE 3-3-1

	Identification Number	Reverse Resistance	Forward Resistance
1			
2			
3			
4			
5			

Summary

A diode can be tested by checking the forward-to-reverse resistance ratio with an ohmmeter. The resistance ratio indicates the diode's ability to pass current in one direction and block current in the other direction. If the positive lead is connected to the cathode and the negative lead to the anode, the diode is reverse biased. Little current should flow and the ohmmeter should indicate a high resistance. When the positive lead is connected to the anode and the negative lead to the cathode, the diode is forward biased. Current flows through the diode and the ohmmeter indicates a low resistance.

If the diode showed both low forward and low reverse resistance, it was probably shorted. If the diode measured both high forward and high reverse resistance, it was probably opened.

The leads of an unmarked diode can also be determined with the ohmmeter. When the ohmmeter reads low resistance, the positive lead is on the anode and the negative lead is on the cathode.

Notes

LAB 3-4

Zener Diodes

Name

Course

Date Due

Objective

The student should be able to identify the terminology and characteristics associated with zener diodes.

Reference

Chapter 21, pages 192-196

Materials Required

None

Equipment Required

None

Notes

Definitions

Define the following terms in complete sentences.

1. Zener Diode

2. Zener Region

3. Breakdown Voltage

4. Derating Factor

5. Maximum Zener Current

6. Positive Zener Voltage Temperature Coefficient

7. Negative Zener Voltage Temperature Coefficient

8. Zener Voltage Rating

Directions

Draw the required circuits and diagrams.

1. Draw the symbol for a zener diode and label all parts.

2. Draw a zener diode regulator circuit using schematic symbols.

LAB 3-4

Zener Diodes

Notes

LAB 3-5

Testing Zener Diodes

Name
Course
Date Due

Objective

The student should be able to accurately test zener diodes.

Reference

Chapter 21, page 195-196

Materials Required

5 Assorted Zener Diodes

Equipment Required

Analog Ohmmeter
or
Digital Ohmmeter with Diode Testing Capabilities

Notes

Safety Precautions

Some ohmmeters use a battery voltage, which can destroy a PN junction.

Procedure

As each step is completed, check it off.

1. Identify the leads with each zener diode package, cathode and anode.
2. Record the identification number of each diode on Table 3-5-1.
3. Select the first zener diode listed on Table 3-5-1.
4. Select the ohmmeter range as follows:

   Digital Ohmmeter - 20 kΩ unless meter has a special diode test position

   Analog Ohmmeter - R x 100
5. Connect the positive or red lead of the ohmmeter to the cathode of the diode and the negative or black lead of the ohmmeter to the anode of the diode.
6. Record the resistance reading on Table 3-5-1.
7. Reverse the leads to the diode, positive to anode and negative to cathode, and record the results on Table 3-5-1.
8. Repeat steps 4 - 7 for the remaining zener diodes listed on Table 3-5-1.

FIGURE 3-5-1

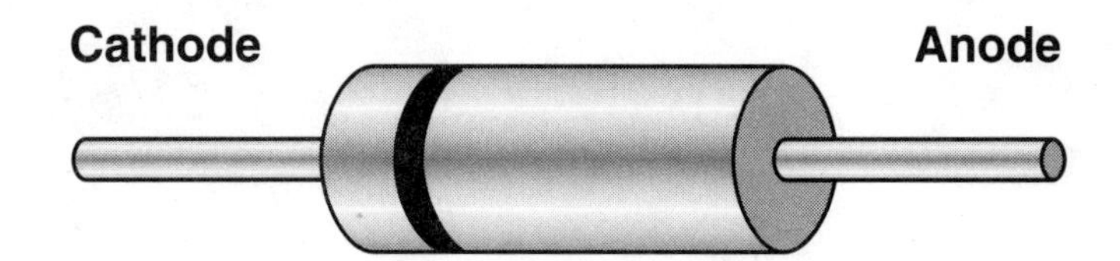

Directions

Complete Table 3-5-1 using steps 4 - 7 on page 109.

TABLE 3-5-1

	Identification Number	Reverse Resistance	Forward Resistance
1			
2			
3			
4			
5			

Summary

A zener diode can be tested for opens, shorts, or leakage with an ohmmeter. The resistance ratio indicates the diode's ability to pass current in one direction and block current in the other direction, the same as with PN junction diodes. If the positive lead is connected to the cathode and the negative lead to the anode, the diode is reverse biased. Little current should flow and the ohmmeter should indicate a high resistance. When the positive lead is connected to the anode and the negative lead to the cathode, the diode is forward biased. Current flows through the diode and the ohmmeter indicates a low resistance.

The resistance test performed does not provide information on whether the zener diode is regulating at the rated value. For that, a regulation test must be performed as identified on page 195 in the textbook.

LAB 3-5

Testing Zener Diodes

Notes

LAB 3-6

Bipolar Transistors

Name

Course

Date Due

Objective

The student should be able to identify the terminology and characteristics associated with bipolar transistors.

Reference

Chapter 22, pages 197-204

Materials Required

None

Equipment Required

None

Notes

Definitions

Define the following terms in complete sentences.

1. Bipolar Transistor

2. Base

3. Emitter

4. NPN Transistor

5. PNP Transistor

Directions

Draw the required circuits and diagrams.

1. Draw the schematic symbol for an NPN transistor and label all parts.

2. Draw the schematic symbol for an PNP transistor and label all parts.

Directions

Draw the required circuits and diagrams.

3. Draw the schematic symbol for a PNP transistor and label all parts.

4. Draw an NPN transistor that is properly biased (use schematic symbols).

5. Draw a PNP transistor that is properly biased (use schematic symbols).

LAB 3-6

Bipolar Transistors

Notes

LAB 3-7

Testing Bipolar Transistors

Name

Course

Date Due

Objective

The student should be able to accurately test bipolar transistors.

Reference

Chapter 22, pages 201-202

Materials Required

5 Assorted Transistors

Equipment Required

Analog Ohmmeter

or

Digital Ohmmeter with Diode Testing Capabilities

Notes

Safety Precautions

Some ohmmeters use a battery voltage, which can destroy a PN junction.

Procedure

As each step is completed, check it off.

1. Identify the leads with each transistor package, emitter, base, and collector.
2. Record the identification number of each diode on Table 3-7-1.
3. Select the first transistor listed on Table 3-7-1.
4. Select the ohmmeter range as follows:

   Digital Ohmmeter - 20 kΩ unless meter has a special diode test position

   Analog Ohmmeter - R x 100

5. Connect the positive or red lead of the ohmmeter to the emitter of the transistor and the negative or black lead of the ohmmeter to the base of the transistor.
6. Record the resistance reading on Table 3-7-1.
7. Reverse the leads to the transistor positive to base and negative to emitter, and record the results on Table 3-7-1.
8. Repeat steps 4 - 7 for the remaining zener diodes listed on Table 3-7-1.

FIGURE 3-7-1

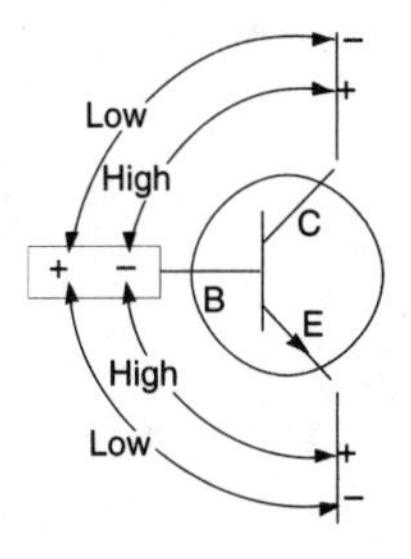

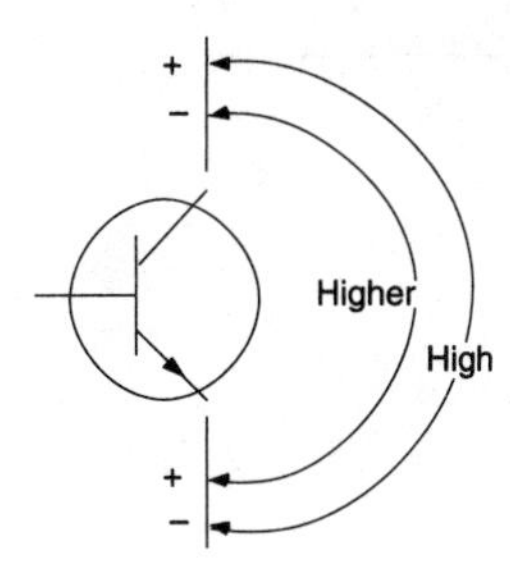

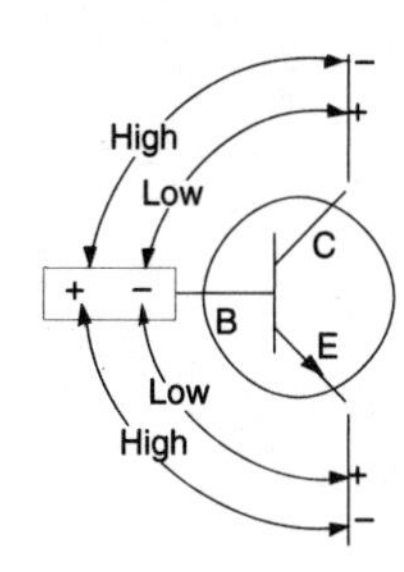

Directions

Complete Table 3-7-1 using steps 4 - 7 on page 113.

TABLE 3-7-1

E — B Junction

	Identification Number	Reverse Resistance	Forward Resistance
1			
2			
3			
4			
5			

C — B Junction

	Identification Number	Reverse Resistance	Forward Resistance
1			
2			
3			
4			
5			

E — C Junction

	Identification Number	Reverse Resistance	Forward Resistance
1			
2			
3			
4			
5			

Notes

LAB 3-8

Field Effect Transistors (FETs)

Name	
Course	
Date Due	

Objective

The student should be able to identify the terminology and characteristics associated with field effect transistors (FETs).

Reference

Chapter 23, pages 205-214

Materials Required

None

Equipment Required

None

Notes

Definitions

Define the following terms in complete sentences.

1. Junction Field Effect Transistor

2. Substrate

3. Channel

4. N-channel JFET

5. P-channel JFET

6. Source

7. Gate

8. Drain

9. Drain Current

10. Pinch-off Voltage

11. MOSFET

12. Depletion Mode MOSFET

13. Enhancement Mode MOSFET

Questions

Answer the following questions in complete sentences.

1. Draw the symbol for an N-channel JFET and label all parts.

2. Draw the symbol for a P-channel JFET and label all parts.

3. Draw an N-channel JFET that is properly biased (use schematic symbols).

4. Draw the symbol for an N-channel depletion mode JFET.

5. Draw the symbol for a P-channel depletion mode JFET.

6. Draw an N-channel depletion MOSFET that is properly biased (use schematic symbols).

7. Draw a P-channel enhancement MOSFET that is properly biased (use schematic symbols).

Notes

LAB 3-9

Name

Course

Date Due

Testing FETs

Objective

The student should be able to accurately test field effect transistors

Reference

Chapter 23, pages 212-213

Materials Required

N-Channel JFET
P-Channel JFET
N-Channel Depletion MOSFET

Equipment Required

Analog Ohmmeter
or
Digital Ohmmeter with Diode Testing Capabilities

Notes

Safety Precautions

MOSFETs require special handling because they are easily damaged by static electricity.

Procedure

As each step is completed, check it off.

1. Identify the leads with each JFET package, source gate, and drain.
2. Record the identification number of each JFET on Table 3-9-1.
3. Select the N-Channel JFET listed on Table 3-9-1.
4. Select the ohmmeter range as follows:

   Digital Ohmmeter - 20 kΩ unless meter has a special diode test position

   Analog Ohmmeter - R x 100

5. Connect the positive or red lead of the ohmmeter to the gate of the JFET and the negative or black lead of the ohmmeter to the source or drain of the JFET.

   ***NOTE**: Because the drain and source are connected, only one side needs to be tested.*

6. Record the resistance reading on Table 3-9-1.
7. Reverse the ohmmeter leads to the gate and to the source or drain and record the results on Table 3-9-1.
8. Repeat steps 4 - 7 for the P-channel JFET and MOSFET listed on Table 3-9-1.

Directions

Complete Table 3-9-1 using steps 4 - 7 on page 117.

TABLE 3-9-1

	Gate-Source		
Type	**Identification Number**	**Forward Resistance**	**Reverse Resistance**
1 • N–Channel JFET			
2 • P–Channel JFET			
3 • MOSFET			

Summary

Testing a field effect transistor is more complicated than testing a normal transistor. This is because of the number of different types of JFETs and MOSFETs.

MOSFETs require special handling instructions because they can be easily damaged by static electricity. MOSFETs have extremely high input resistance because of the insulated gate. The forward and reverse resistance should be checked with a low-voltage ohmmeter on its highest scale.

The ohmmeter should have registered an infinite resistance in both the forward and reverse resistance test between the gate and source or drain. A lower reading indicates a breakdown of the insulation between the gate and source or drain.

Notes

LAB 3-10

Thyristors

Name

Course

Date Due

Objective

The student should be able to identify the terminology and characteristics associated with thyristors.

Reference

Chapter 24, pages 215-223

Materials Required

None

Equipment Required

None

Notes

Definitions

Define the following terms in complete sentences.

1. Silicon Controlled Rectifier

2. Regenerative Feedback

3. Cathode

4. Anode

5. Gate

6. TRIAC

7. Main Terminal 1

8. Main Terminal 2

9. DIAC

Questions

Answer the following questions in complete sentences.

1. Draw the schematic symbol for an SCR and label all parts.

2. Draw the schematic symbol for a TRIAC and label all parts.

3. Draw the schematic symbol for a DIAC.

LAB 3-10

Thyristors

Notes

LAB 3-11

Testing SCRs

Name
Course
Date Due

Objective

The student should be able to accurately test silicon controlled rectifiers.

Reference

Chapter 24, pages 221-222

Materials Required

5 Assorted SCRs

Equipment Required

Analog Ohmmeter
or
Digital Ohmmeter with
Diode Testing Capabilities

Notes

Procedure

As each step is completed, check it off.

1. Identify the leads with each SCR package, anode, cathode, and gate.
2. Record the identification number of each SCR on Table 3-11-1.
3. Select the first SCR listed on Table 3-11-1.
4. Select the ohmmeter range as follows:

   Digital Ohmmeter - 20 kΩ unless meter has a special diode test position

   Analog Ohmmeter - R x 100
5. Connect the positive or red lead of the ohmmeter to the cathode of the SCR and the negative or black lead of the ohmmeter to the anode of the SCR.
6. Record the resistance reading on Table 3-11-1.
7. Reverse the ohmmeter leads, positive to the anode and negative to the cathode and record the results on Table 3-11-1.
8. With the ohmmeter leads connected as in step 7, short the gate to the anode (touch the gate leads to the anode lead) and record the results on Table 3-11-1.
9. Remove the short between the gate and the anode. The resistance should stay low.
10. Remove the ohmmeter leads from the SCR and repeat the test.
11. Repeat steps 4 - 10 for the remaining SCRs listed on Table 3-11-1.

Directions

Complete Table 3-11-1 using steps 4 - 10 on page 121.

TABLE 3-11-1

Cathode - Anode

	Identification Number	Reverse Resistance	Forward Resistance	Resistance with Gate Short to Anode
1				
2				
3				
4				
5				

Summary

Before the gate was shorted to the anode, the resistance should have exceeded 1 MΩ in both directions. When the gate was shorted to the anode, the resistance should have dropped to less than 1 MΩ. When the short was removed between the anode and the gate, the resistance should have stayed low. However, if the resistance range used was high, the resistance should have returned to above 1 MΩ. In the higher resistance ranges, the ohmmeter does not supply enough current to keep the gate latched (turned on) when the short is removed.

The ohmmeter can detect the majority of defective thyristors. It cannot detect marginal or voltage-sensitive devices. However, it can give a good indication of the condition of the thyristor.

LAB 3-11

Testing SCRs

Notes

LAB 3-12

Integrated Circuits

Name

Course

Date Due

Objective

The student should be able to identify the terminology and characteristics associated with integrated circuits.

Reference

Chapter 25, pages 225-229

Materials Required

None

Equipment Required

None

Notes

Definitions

Define the following terms in complete sentences.

1. Integrated Circuit

2. Monolithic

3. Thin Film

4. Thick Film

5. Hybrid

6. Yield

7. Evaporation Process

8. Sputtering Process

9. DIP

10. Small-scale Integration

11. Medium-scale Integration

12. Large-scale Integration

13. Very Large-scale Integration

Questions

Answer the following questions in complete sentences.

1. What are the advantages of using ICs?

2. What are the disadvantages of using ICs?

LAB 3-12

Integrated Circuits

Notes

LAB 3-13

IC Identification

Name
Course
Date Due

Objective

The student should be able to accurately identify an IC's type and the DIP package pinout.

Reference

Chapter 25, pages 225-229

Materials Required

5 Assorted ICs

Equipment Required

IC Reference Manual

Notes

Procedure

As each step is completed, check it off.

1. Break down the data listed on the IC page: manufacturer, IC identification number, and date code.
2. Record the information on Table 3-13-1.
3. Select the first IC listed on Table 3-13-1.
4. Look up the identification number in the reference manual.
5. Record the function of the IC, i.e., quad AND gate, hex inverter, etc.
6. Label the IC's pins on Figure 3-13-1.
7. Repeat steps 4 - 6 for each of the remaining ICs on Table 3-11-1.

FIGURE 3-13-1

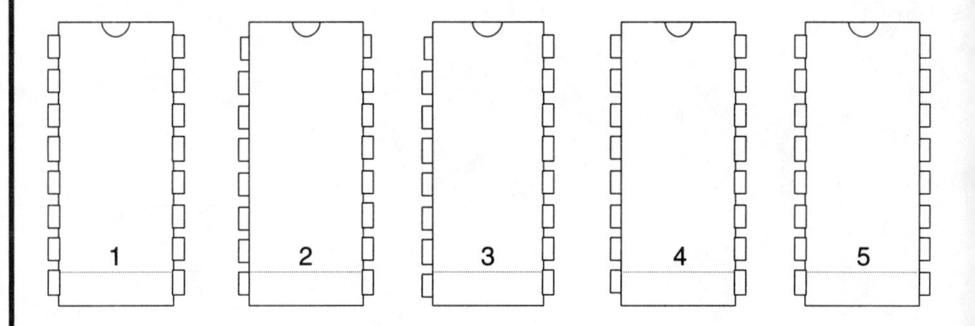

Directions

Complete Table 3-13-1 using steps 4 - 6 on page 125.

TABLE 3-13-1

	Identification Number	IC Family Type	IC Function
1			
2			
3			
4			
5			

Summary

The information recorded on the IC package tells a lot about the IC. From the identification number, the purpose of the IC is identified. The date code tells what year and week the IC was manufactured. This is useful if a bad batch was discovered by the manufacturer. Each manufacturer has in-house codes for identification as well as the standard coding. TTL ICs are identified with a 54XX or a 74XX number. CMOS ICs are identified with a 40XX number. The manufacturer reference manuals are essential to identifying the IC's function.

Notes

LAB 3-14

Optoelectric Devices

Name

Course

Date Due

Objective

The student should be able to identify the terminology associated with optoelectric devices.

Reference

Chapter 26, pages 230-237

Materials Required

None

Equipment Required

None

Notes

Definitions

Define the following terms in complete sentences.

1. Light

2. Photoconductive Cell

3. Photovoltaic Cell

4. Photodiode

5. PIN Photodiode

6. Phototransistor

7. Light-Emitting Diode

8. Optical Coupler

Questions

Answer the following questions in complete sentences.

1. Draw the schematic symbol for a photo cell.

2. Draw and label the schematic symbol for a photovoltaic cell.

3. Draw and label the schematic symbol for a photodiode.

4. Draw and label the schematic symbol for a phototransistor.

5. Draw and label the schematic symbol for an LED.

Notes

LAB 3-15

Testing LEDs

Name

Course

Date Due

Objective

The student should be able to accurately test an LED.

Reference

Chapter 26, pages 230-237

Materials Required

5 Assorted LEDs; Jumbo and Miniature

Equipment Required

Analog Ohmmeter

or

Digital Ohmmeter with Diode Testing Capabilities

LED Reference Manual

Notes

Safety Precautions

Never connect an LED to a power source to test it without connecting a resistor in series with it.

Procedure

As each step is completed, check it off.

1. Identify the leads of the LEDs and label them, anode and cathode.
2. Record the LEDs on Table 3-15-1. Some LEDs have part numbers printed on them.
3. Select the first LED listed on Table 3-15-1.
4. Select the ohmmeter range as follows:

   Digital Ohmmeter - 20 kΩ unless meter has a special diode test position

   Analog Ohmmeter - R x100
5. Connect the positive or red lead of the ohmmeter to the cathode and connect the negative or black lead of the ohmmeter to the anode of the LED.
6. Record the resistance readings on Table 3-15-1.
7. Reverse the leads of the ohmmeter, positive to anode and negative to cathode, and record the resistance reading on Table 3-15-1.
8. Repeat steps 4 - 7 for the remaining LEDs listed on Table 3-15-1.

Directions

Complete Table 3-15-1 using steps 4 - 7 on page 129.

TABLE 3-15-1

	Identification Number	Reverse Resistance	Forward Resistance
1			
2			
3			
4			
5			

Summary

The LED is really a diode that emits light. The cathode is identified as the lead closest to the flat side. When the positive lead was on the anode and the negative lead was on the cathode, the LED was forward biased and conducted current flow. The resistance was low. When the leads were reversed, the LED was reverse biased and did not support current flow. The resistance reading was high.

LEDs can be easily damaged by an excessive amount of current or voltage. A series resistor must be connected to limit current flow. Never connect an LED to a power source to test it without connecting a resistor in series with it.

LAB 3-15

Testing LEDs

Notes

LAB 4-1

Power Supplies

Name

Course

Date Due

Objective
The student should be able to identify the terminology associated with power supplies.

Reference
Chapter 27, pages 241-256

Materials Required
None

Equipment Required
None

Notes

Definitions
Define the following terms in complete sentences.

1. Transformer
2. Half-Wave Rectifier
3. Full-Wave Rectifier
4. Bridge Rectifier
5. Ripple Frequency
6. Filter
7. Voltage Regulator
8. Shunt Regulator
9. Series Regulator
10. Voltage Multiplier
11. Voltage Doubler
12. Voltage Tripler

Definitions

Define the following terms in complete sentences.

13. Over-voltage Protection Circuit

14. Fuse

15. Circuit Breaker

Questions.

Answer the following questions in complete sentences.

1. Draw and label the blocks of a power supply.

2. Draw the three basic types of rectifier circuits.

Notes

LAB 4-2

Power Supply Rectifiers

Name

Course

Date Due

Objective

The student should be able to determine the difference between half-wave, full-wave, and bridge rectifiers.

Reference

Chapter 27, pages 235-245

Materials Required

4, 1N4003 Silicon Diodes

4700 Ω, 1/2 W Resistor
Yellow- Violet - Red

12 V AC Transformer

Equipment Required

Protoboard

Voltmeter with Leads

Oscilloscope

Notes

Procedure

As each step is completed, check it off.

1. Connect the circuit as shown in Figure 4-2-1.
2. Turn on the oscilloscope and adjust for a proper display.
3. Apply power to the transformer.
4. Connect the oscilloscope ground to point C and the probe to point A.
5. Record the results on Figure 4-2-2.
6. Connect the oscilloscope probe to point B and record the results on Figure 4-2-3.
7. Turn the power off.
8. Connect the circuit shown in Figure 4-2-4.
9. Apply power to the circuit.
10. Connect the oscilloscope ground to point B and the probe to point A.
11. Record the waveform observed on Figure 4-2-5.
12. Turn the power off.
13. Construct the circuit shown in Figure 4-2-6.
14. Connect the oscilloscope probe to point 8 and record the results on Figure 4-2-3.
15. Apply power to the circuit. Record the waveform observed on Figure 4-2-7.
16. Turn the power off.

Directions

Complete Figures 4-2-2 and 4-2-3 using steps 1 - 7 on page 133.

FIGURE 4-2-1

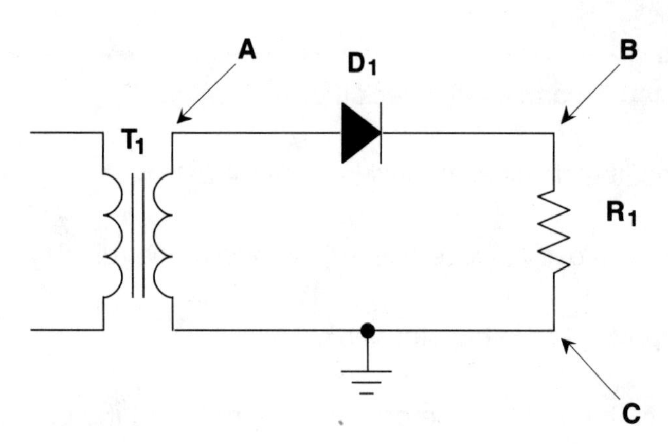

FIGURE 4-2-2

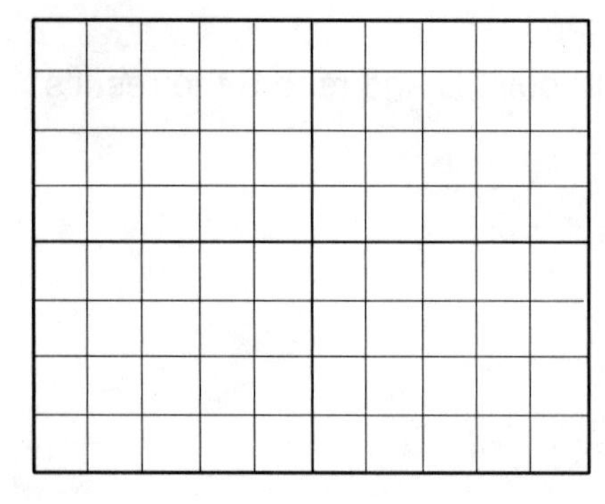

Scope Setting

Volts/cm
Time/cm

Probe Setting

Measured Voltage
Across What Points

FIGURE 4-2-3

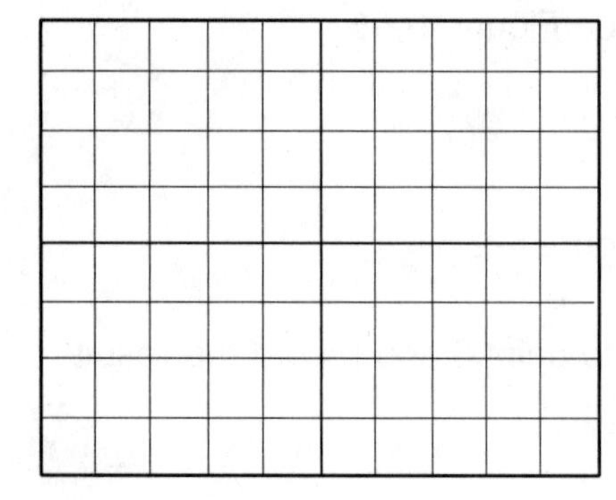

Scope Setting

Volts/cm
Time/cm

Probe Setting

Measured Voltage
Across What Points

LAB 4-2

Power Supply Rectifiers

Notes

LAB 4-2

Power Supply Rectifiers

Name
Course
Date Due

Notes

Directions

Complete Figure 4-2-5 using steps 8 - 12 on page 133.

FIGURE 4-2-4

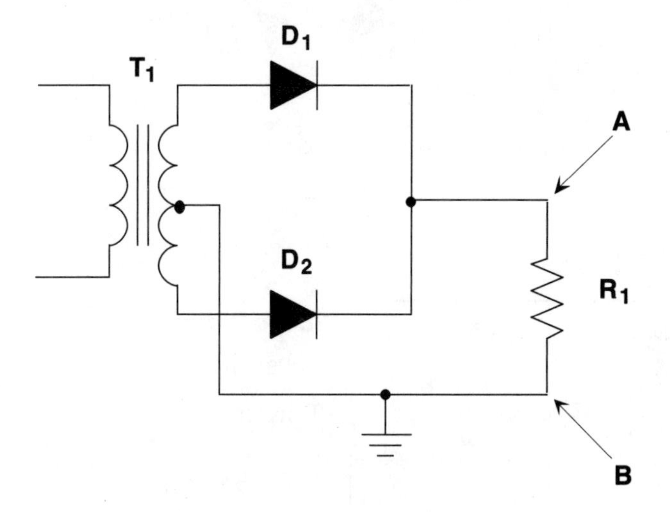

FIGURE 4-2-5

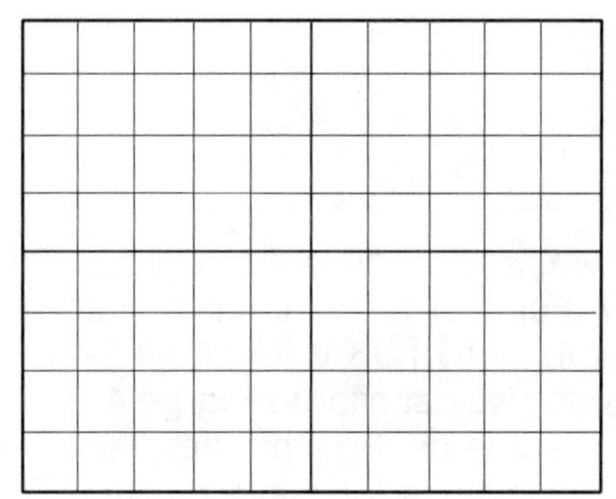

Scope Setting

Volts/cm
Time/cm

Probe Setting

Measured Voltage
Across What Points

LAB 4-2

Power Supply Rectifiers

Directions

Complete Figure 4-2-7 using steps 13 - 16 on page 133.

FIGURE 4-2-6

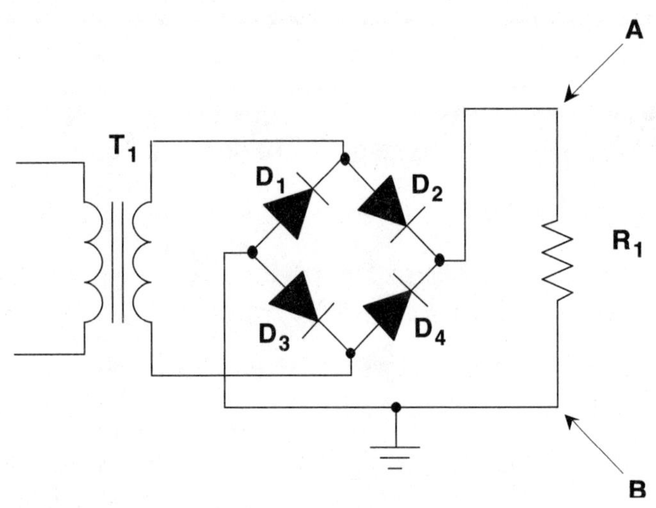

FIGURE 4-2-7

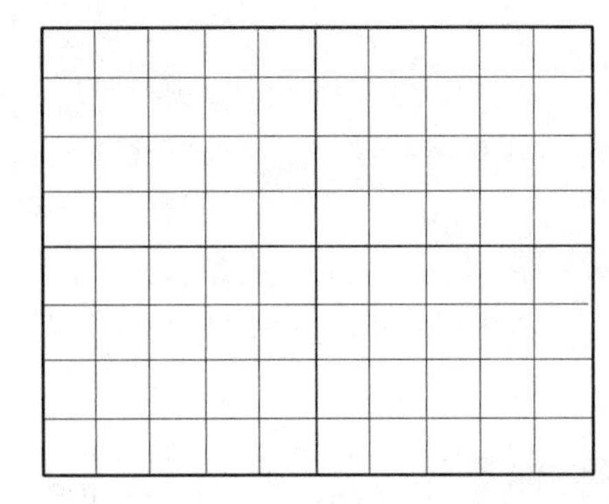

Scope Setting

Volts/cm	
Time/cm	

Probe Setting

Measured Voltage	
Across What Points	

FIGURE 4-2-8

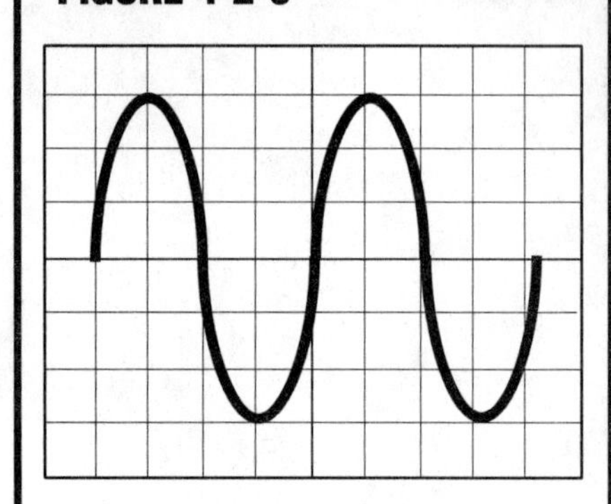

FIGURE 4-2-9

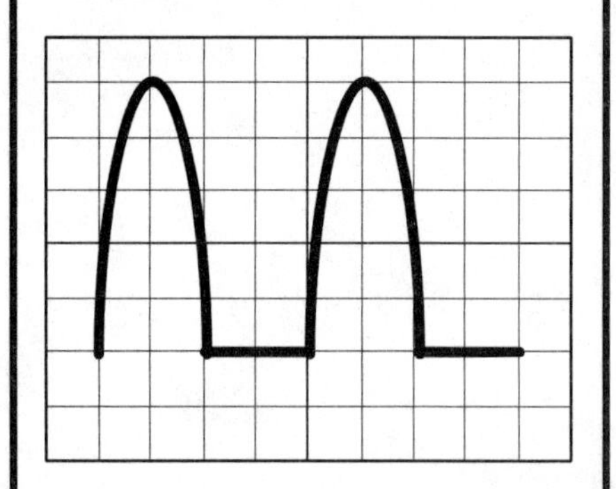

FIGURE 4-2-10

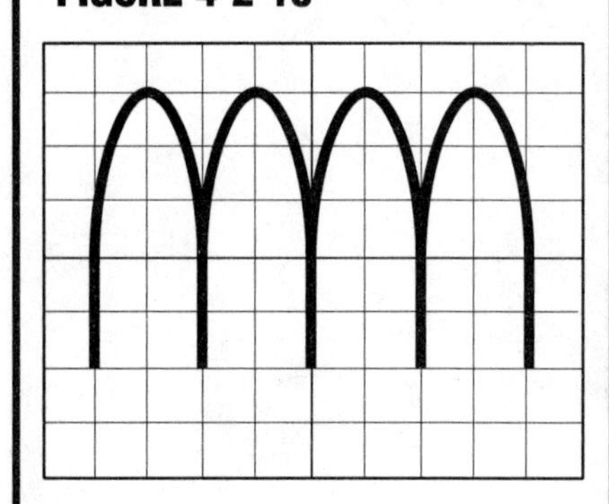

Summary

The circuit constructed in steps 1 - 7 was a half-wave rectifier. The AC input signal has a period of 0.167 s or 16.7 ms. The input voltage was a sine wave with a peak-to-peak voltage of 96 V (Figure 4-2-8). The voltage observed across the resistor was similar to Figure 4-2-9. The output frequency has a period the same as the input frequency.

The circuit constructed in steps 8 - 12 was a full-wave rectifier. The output frequency is twice that of the input frequency. This makes it easier to remove the ripple frequency in a power supply. The recorded results should be the same as shown in Figure 4-2-10.

The circuit constructed in steps 13 -16 was a bridge rectifier. Like the full-wave it also has a ripple frequency of 120 Hz (twice the input frequency) as shown in Figure 4-2-11. Therefore, the bridge rectifier can also be considered a full-wave rectifier. Notice the output voltage was higher than either the half-wave rectifier or the full-wave rectifier.

LAB 4-3

Power Supply Filters

Name
Course
Date Due

Objective

The student should be able to determine how the size of the capacitive filter affects the ripple amplitude and the output voltage.

Reference

Chapter 27, pages 245-247

Materials Required

4, 1N4003 Silicon Diodes
4700 Ω, 1/2 W Resistor, Yellow- Violet- Red
10 μF Electrolytic Capacitor
100 μF Electrolytic Capacitor
12 V AC Transformer

Equipment Required

Protoboard
Voltmeter with Leads
Oscilloscope with Test Probes

Notes

Procedure

As each step is completed, check it off.

1. Connect the circuit shown in Figure 4-3-1.
2. Turn on the oscilloscope and adjust for proper display. The position of the trace on the oscilloscope represents the 0 V DC reference.
3. Connect the oscilloscope ground to point B and the probe to point A and apply power to the circuit.
4. Draw two cycles of the waveform displayed on Figure 4-3-2 and record the voltage drop across the resistor using the voltmeter. Remove power from the circuit and replace the 10 μF capacitor with the 100 μF capacitor.
5. Apply power to the circuit and draw two cycles of the waveform displayed on Figure 4-3-3 and record the voltage drop across the resistor using the voltmeter.
6. Remove power from the circuit and connect the circuit shown in Figure 4-3-4, using the 10 μF capacitor, with the oscilloscope connected at points A and B.
7. Apply power to the circuit and draw two cycles of the waveform displayed on Figure 4-3-5 and record the voltage drop across the resistor using the voltmeter. Remove power from the circuit and replace the 10 μF capacitor with the 100 μF capacitor.
8. Apply power to the circuit and draw two cycles of the waveform displayed on Figure 4-3-6 and measure the voltage drop across the resistor using the voltmeter.
9. Remove power from the circuit and connect the circuit shown in Figure 4-3-7 with the oscilloscope connected at points A and B.
10. Apply power to the circuit and draw two cycles of the waveform displayed on Figure 4-3-8 and record the voltage drop across the resistor using the voltmeter. Remove power from the circuit and replace the 10 μF capacitor with the 10 μF capacitor.
11. Apply power to the circuit and draw two cycles of the waveform displayed on Figure 4-3-9 and record the voltage drop across the resistor using the voltmeter. Remove power from the circuit.

Directions

Complete Figures 4-3-2 and 4-3-3 using steps 1 - 5 on page 137.

FIGURE 4-3-1

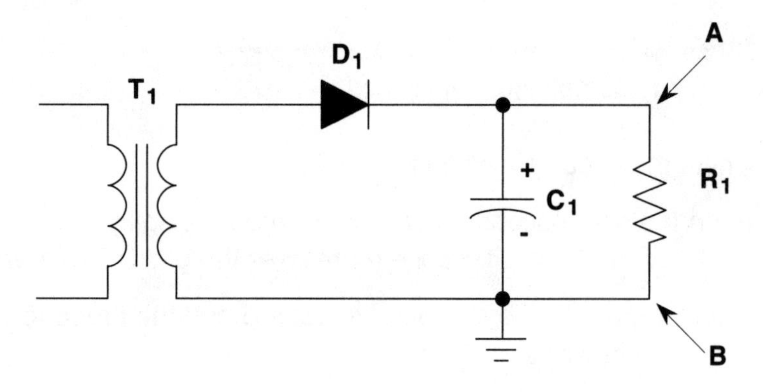

FIGURE 4-3-2

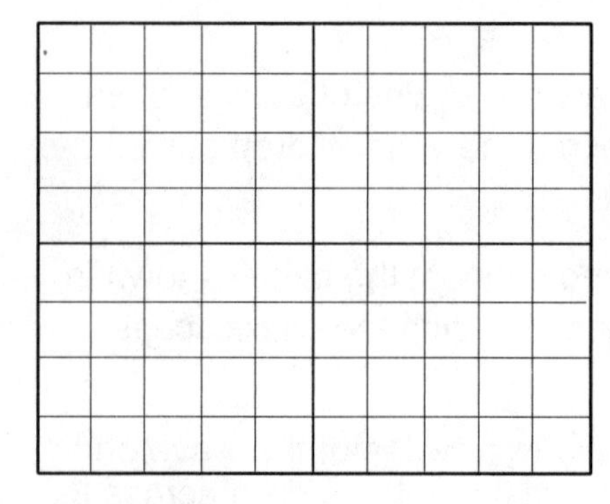

Scope Setting

Volts/cm
Time/cm

Probe Setting

Measured Voltage
Across What Points

FIGURE 4-3-3

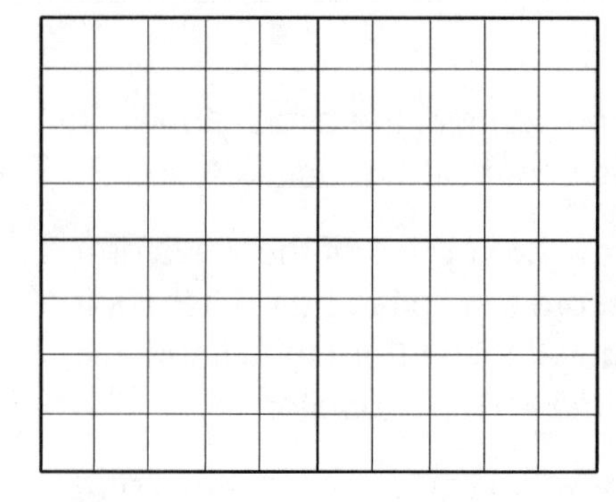

Scope Setting

Volts/cm
Time/cm

Probe Setting

Measured Voltage
Across What Points

LAB 4-3

Power Supply Filters

Notes

LAB 4-3

Power Supply Filters

Name	
Course	
Date Due	

Notes

Directions

Complete Figures 4-3-5 and 4-3-6 using steps 6 - 8 on page 137.

FIGURE 4-3-4

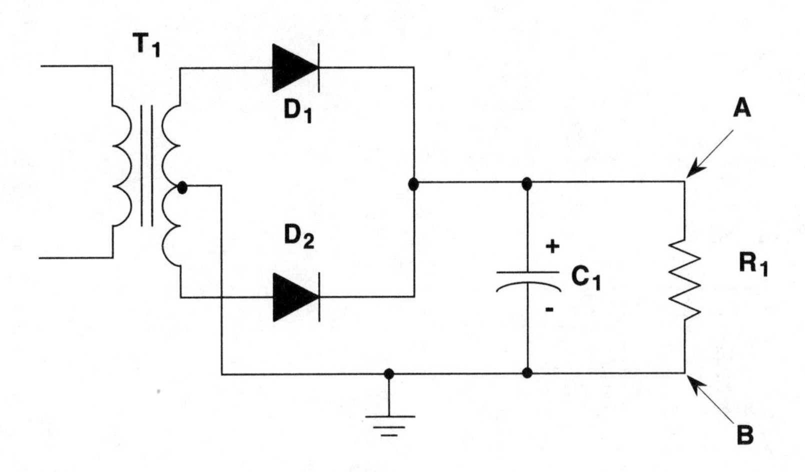

FIGURE 4-3-5

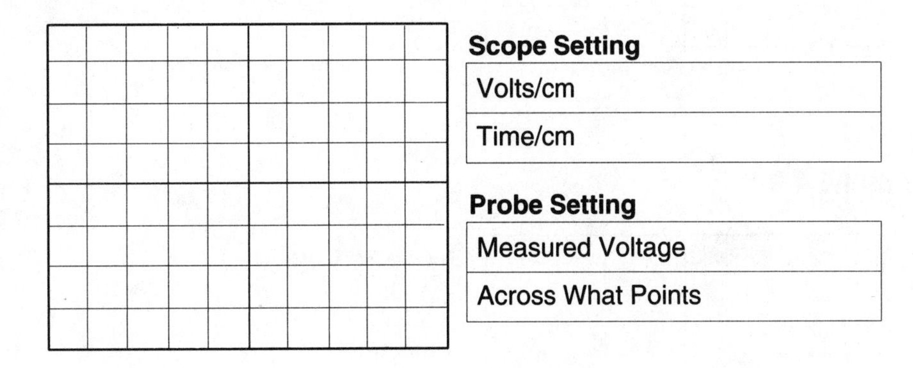

FIGURE 4-3-6

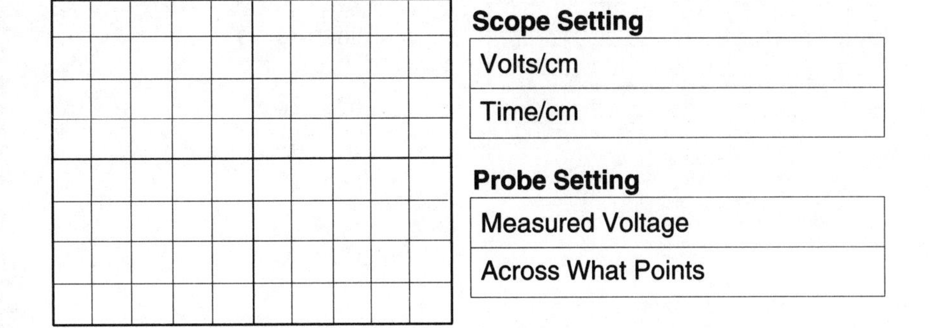

Directions

Complete Figures 4-3-8 and 4-3-9 using steps 9 - 11 on page 137.

FIGURE 4-3-7

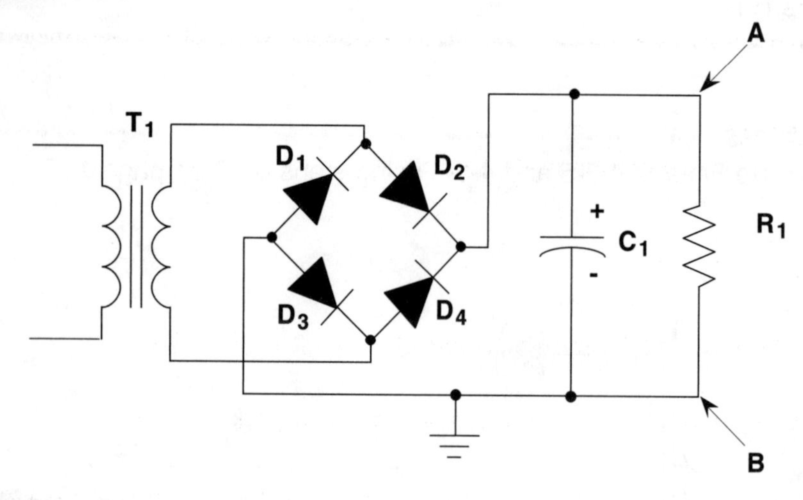

FIGURE 4-3-8

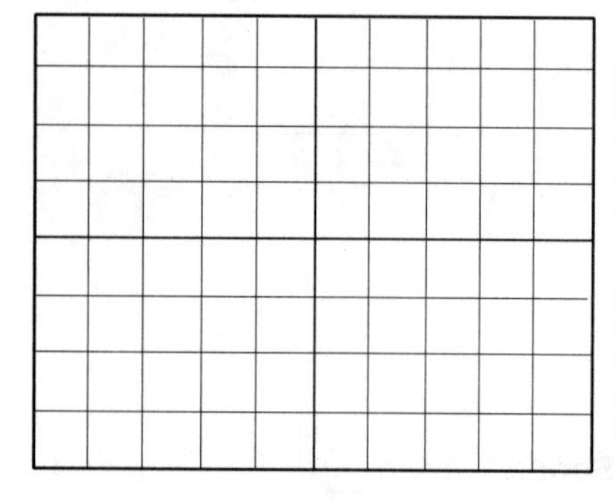

Scope Setting

Volts/cm	
Time/cm	

Probe Setting

Measured Voltage	
Across What Points	

FIGURE 4-3-9

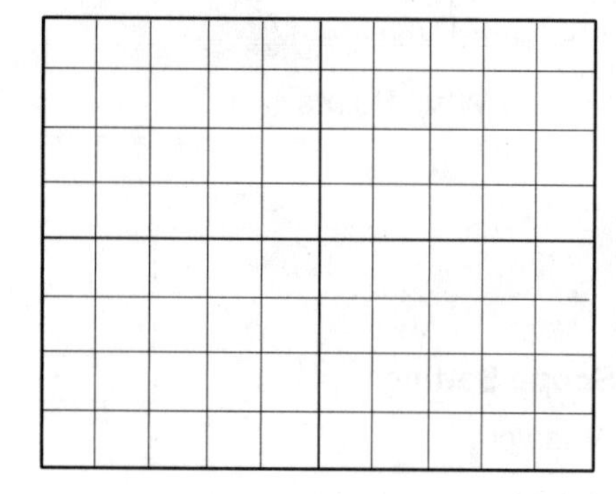

Scope Setting

Volts/cm	
Time/cm	

Probe Setting

Measured Voltage	
Across What Points	

LAB 4-3

Power Supply Filters

Summary

The circuit constructed in steps 1 - 5 was a half-wave rectifier. With the addition of the 10 μF capacitor, the amplitude of the ripple frequency decreased to an unfiltered signal. When the 100 μF capacitor was used in place of the 10 μF capacitor, the ripple frequency decreased even further. The DC level appeared as almost a solid level. There was also a slight increase in the DC output voltage.

The circuit constructed in steps 6 - 8 was a full-wave rectifier. The ripple frequency was twice the ripple frequency of the half-wave rectifier. Notice how much easier it was to filter the higher ripple frequency to produce a DC level. The voltage measured with the DC voltmeter was less than the voltage measured for the half-wave rectifier. This decrease in voltage occurred because of the center tap of the transformer. Only one-half of the voltage was available. When the 10 μF capacitor was replaced with the 100 μF capacitor, the ripple frequency was reduced considerably. The DC level of the output voltage also increased with the change of capacitors.

The circuit constructed in steps 9 - 11 was a bridge rectifier. The ripple frequency was 120 Hz, the same as the full-wave rectifier. The voltage measured across the resistor was higher than either the half-wave or full-wave rectifier.

LAB 4-4

Shunt Voltage Regulation

Name

Course

Date Due

Objective

The student should be able to determine the operation and characteristics of a shunt voltage regulator.

Reference

Chapter 27, pages 247-251

Materials Required

150 Ω, 1/2 W Resistor
620 Ω, 1/2 W Resistor
100 Ω, 1/2 W Potentiometer
6.2 V, 1 W Zener Diode

Equipment Required

Protoboard
Power Supply
Voltmeter

Notes

Procedure

As each step is completed, check it off.

1. Connect the circuit shown in Figure 4-4-1.
2. Using the voltmeter, adjust the power supply for +8 V.
3. Adjust the 100 kΩ potentiometer for its minimum value.
4. Measure the value of the output voltage with the voltmeter. Record the value on Table 4-4-1.
5. Readjust the input voltage for +12 V.
6. Measure the value of the output voltage and record the results on Table 4-4-1.
7. Readjust the input voltage for +10 V.
8. Measure the value of the output voltage and record the results on Table 4-4-1.
9. Adjust the 100 kΩ potentiometer to its maximum value. Record the output voltage on Table 4-4-1.

TABLE 4-4-1

Line #	E_{In}	R_3	E_{Out}
1	8 V	Min	
2	12 V	Min	
3	10 V	Min	
4	10 V	Max	

Directions

Use Figure 4-4-1 for steps 1 - 9 on page 141.

FIGURE 4-4-1

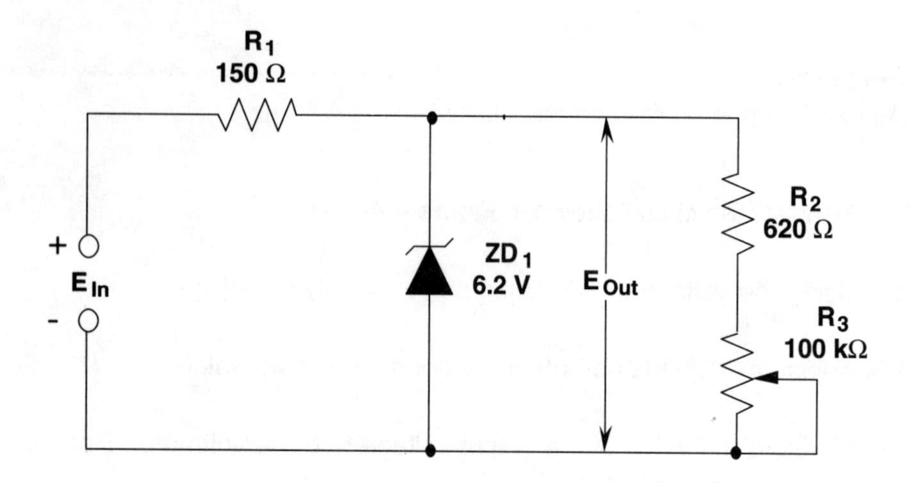

Summary

The circuit constructed was a simple zener diode regulator. The zener diode had a zener voltage of 6.2 V. It held the output voltage across the load to 6.2 V. Changing the input voltage from 8 to 12 V resulted in only a slight change in the output voltage across the load. Initially, the load resistance was set for 620 Ω. The current through the load was approximately 10 mA. This was the full load current. When the potentiometer was set to its maximum value, the load current decreased to approximately 62 μA. The output voltage across the load remained at 6.2 V. The disadvantage with zener diode regulators is their low current handling capabilities.

LAB 4-4

Shunt Voltage Regulation

Notes

LAB 4-5

Series Voltage Regulation: Emitter-Follower Regulator

Name

Course

Date Due

Objective

The student should be able to determine the operation and characteristics of a series emitter-follower voltage regulator.

Reference

Chapter 27, pages 247-251

Materials Required

100 Ω, 1/2 W Resistor
1800 Ω, 1/2 W Resistor
1 kΩ, 2 W Potentiometer
6.2 V Zener Diode
2N3055 NPN Transistor

Equipment Required

Protoboard
Power Supply
Voltmeter with Leads

Notes

Procedure

As each step is completed, check it off.

1. Connect the circuit shown in Figure 4-5-1.
2. Using the voltmeter, adjust the power supply for +10 V.
3. Adjust potentiometer R_3 for its maximum value.
4. Measure the value of the voltage across the zener diode. Record the value on Table 4-5-1.
5. Measure and record the voltage across the base-emitter junction of the transistor on Table 4-5-2.
6. Measure and record the output voltage on Table 4-5-2.
7. Does the output voltage equal the voltage dropped across the zener diode minus the base-emitter junction voltage of the transistor? Record the answer on Table 4-5-2.
8. Complete Table 4-5-1 by adjusting the input voltage to the values indicated for potentiometer R_3 set to its maximum value.
9. Set potentiometer R_3 to its minimum value.
10. Adjust the input voltage for +10 V.
11. Measure and record the output voltage on Table 4-5-1.
12. Does the output voltage remain constant for changes in the input voltage? Record the answer on Table 4-5-2.

Directions

Use Figure 4-5-1 for steps 1 - 12 on page 143.

FIGURE 4-5-1

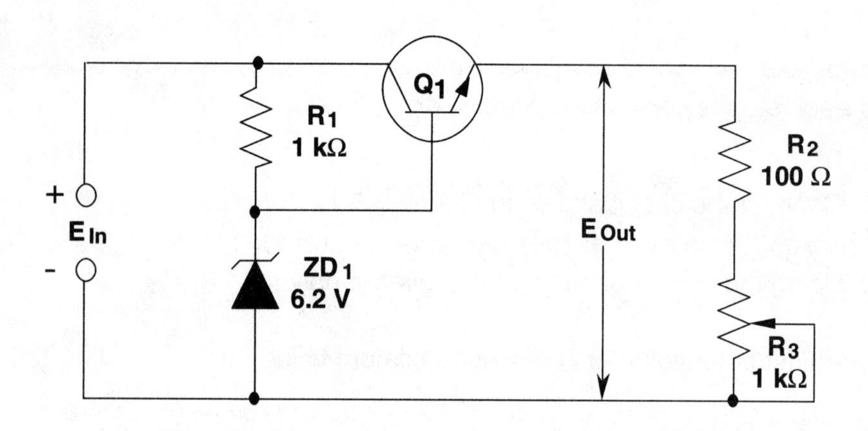

TABLE 4-5-1

Step	E_{IN}	R_3	E_{OUT}
1	10 V	Min	
2	10 V	Max	
3	13 V	Min	
4	13 V	Max	
5	16 V	Min	
6	16 V	Max	

TABLE 4-5-2

Step	Question	Answer
4	E_Z =	
5	E_{B-E} =	
6	E_{OUT} =	
7	Does E_{OUT} = E_{B-E} ?	
12	Does E_{OUT} remain constant?	

LAB 4-5

Series Voltage Regulation: Emitter-Follower Regulator

Summary

The emitter-follower regulator allows a zener diode to regulate a voltage at a higher current level than the zener diode regulator. The output voltage is always less than the zener diode voltage. The output voltage is equal to the zener voltage minus the voltage drop across the base-emitter junction of the transistor. The output voltage remains constant for both an input voltage increase and a load change. The output voltage did not increase more than 0.2 V. A disadvantage of the emitter-follower regulator is that the output voltage is always less than the zener voltage.

LAB 4-6

Name

Course

Date Due

Series Voltage Regulation: Series Feedback Regulator

Objective

The student should be able to determine the operation and characteristics of a series emitter-follower voltage regulator.

Reference

Chapter 27, pages 247-251

Materials Required

10 Ω, 1/2 W Resistor
330 Ω, 1/2 W Resistor
2, 1 kΩ, 1/2 W Resistor
47kΩ, 1/2 W Resistor
1kΩ, 2 W Potentiometer
2.7 V Zener Diode
3, 2N2222 Silicon Transistors
2N3055 Power Transistor

Equipment Required

Protoboard
Power Supply
Voltmeter with Leads

Notes

Procedure

As each step is completed, check it off.

1. Connect the circuit shown in Figure 4-6-1.
2. Using the voltmeter, adjust the power supply for +10 V.
3. Adjust the output voltage to +5 V using potentiometer R_5.
4. Connect the voltmeter across load resistor R_7.
5. Disconnect resistor R_7. Record the output voltage on Table 4-6-1.
6. Reconnect resistor R_7, readjust the input voltage to +15 V, and record the output voltage on Table 4-6-1.
7. Readjust the input voltage for +12 V.
8. Adjust potentiometer R_5 fully counterclockwise and record the output voltage on Table 4-6-1.
9. Adjust potentiometer R_5 fully clockwise and record the output voltage on Table 4-6-1.
10. Adjust the output voltage for +6 volts using potentiometer R_5.
11. Measure the voltage drop across resistor R_3 and record it on Table 4-6-2.
12. Using Ohm's Law, calculate the current flowing through resistor R_3 and record it on Table 4-6-3.
13. Using a jumper wire, short out resistor R_7 and measure the voltage drop across resistor R_3. Record it on Table 4-6-2.
14. Remove the jumper wire from across resistor R_7.
15. Using Ohm's Law, calculate the current flowing through resistor R_3 with resistor R_7 shorted. Record it on Table 4-6-3.
16. Remove transistor Q_4 from the circuit and measure the voltage drop across resistor R_3. Record it on Table 4-6-2.
17. Using a jumper wire, short out resistor R_7 and measure the voltage drop across resistor R_3. Record it on Table 4-6-2.
18. Remove the jumper wire from across resistor R_7. Disconnect the power from the circuit.
19. Using Ohm's Law, calculate the current flowing through resistor R_3 with resistor R_7 shorted and record it on Table 4-6-3.

LAB 4-6

Series Voltage Regulation: Series Feedback Regulator

Directions

Use Figure 4-6-1 for steps 1 - 19 on page 145.

FIGURE 4-6-1

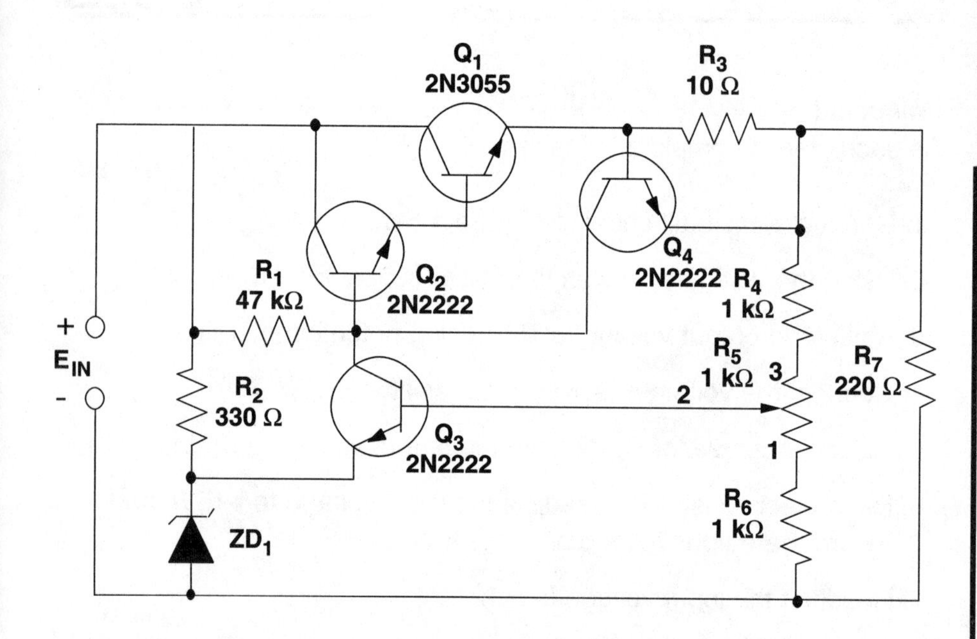

TABLE 4-6-1

Step	E_{OUT}
5	Volts
6	Volts
8	Volts
9	Volts

TABLE 4-6-2

Step	E R_3
11	Volts
13	Volts
16	Volts
17	Volts

TABLE 4-6-3

Step	I R_3
12	mA
15	mA
20	mA

Summary

The circuit constructed is a series feedback regulator. Transistor Q_1 is the series regulator transistor. Transistor Q_1 and transistor Q_2 are connected as a Darlington configuration. Transistor Q_4 is the error detector and amplifier. Transistor Q_4 and resistor R_3 form the current limiting circuit.

Initially, the input voltage was adjusted for 10 V and the output voltage was adjusted for 5 V. When resistor R_7, the load resistor, was disconnected, the current through the regulator dropped to the minimum value. The output voltage still remained at 5 V. The input voltage was then adjusted to 15 V. The output voltage increased slightly. The large change in the input voltage resulted in a very small change in the output voltage. Next, potentiometer R_5 was adjusted fully counterclockwise . The output voltage was about 5.2 V. Then, potentiometer R_5 was adjusted fully clockwise. The output voltage increased to about 9.2 V. Potentiometer R_5 allowed the output voltage to nearly double.

The operation of the current limiting circuit was verified. The voltage drop across resistor R_Q was measured. The current through resistor R_3 was determined using Ohm's law. Initially, the current was about 32 mA when the output voltage was 6 V. When resistor R_7 was shorted, the current increased slightly. The voltage drop across resistor R_3 increased about 0.6 V. This increase in voltage turned on transistor Q_4, decreasing the conduction of transistor Q_1. The maximum current that can flow is limited to about 60 mA, even though resistor R_7 is shorted.

Transistor Q_4 was removed from the circuit. The full-load current remained at 32 mA. When resistor R_7 was shorted, the current increased to over 100 mA. This proved that transistor Q_4 controls the output current flow.

LAB 4-7

Amplifiers

Name

Course

Date Due

Objective

The student should be able to identify the terminology and circuits associated with amplifiers.

Reference

Chapter 28, pages 257-265

Materials Required

None

Equipment Required

None

Notes

Definitions

Define the following terms in complete sentences.

1. Amplifier
2. Amplification
3. Common-base Amplifier
4. Common-emitter Amplifier
5. Common-collector Amplifier
6. Degenerative Feedback
7. Class A Amplifier
8. Class AB Amplifier
9. Class B Amplifier
10. Class C Amplifier
11. Coupling
12. Direct-coupled Amplifier
13. Differential Amplifier
14. Audio Amplifier

Definitions

Define the following terms in complete sentences.

15. Voltage Amplifier

16. Power Amplifier

17. Video Amplifier

18. RF Amplifier

19. Operational Amplifier

20. Summing Amplifier

21. Differential Amplifier

Questions

Answer the following questions in complete sentences.

1. Draw schematic diagrams of the three basic types of amplifiers and label each of them.

2. Draw examples of the four types of coupling techniques.

LAB 4-7

Amplifiers

Notes

LAB 4-8

Common Emitter Amplifier

Name

Course

Date Due

Objective

The student should be able to determine the operation and characteristics of a common emitter amplifier.

Reference

Chapter 28, pages 257-265

Materials Required

1 kΩ, 1/2 W Resistor
1 kΩ, 2 W Potentiometer
4.7 kΩ, 1/2 W Resistor
2, 10 kΩ, 1/2 W Resistor
56 kΩ, 1/2 W Resistor
47 μF Electrolytic Capacitor
2N2222 Transistor

Equipment Required

Protoboard
Voltmeter with Leads
Oscilloscope with Test Probes
Power Supply
Signal or Function Generator with Leads

Notes

Procedure

As each step is completed, check it off.

1. Connect the circuit shown in Figure 4-8-1.
2. Turn on the oscilloscope and adjust for a proper display.
3. Turn on and adjust the power supply for +10 V.
4. Set the signal generator for a 1 KHz signal at 1 V.
5. Connect the signal generator to the input of the circuit; pin 1 on the potentiometer.
6. Connect the oscilloscope to the signal generator output and observe the signal.
7. Turn the potentiometer fully clockwise; the input should be reduced to 0 V.
8. Connect the oscilloscope probe to the collector of the transistor and connect the ground lead to the circuit ground.
9. Turn the potentiometer slowly clockwise. What happens to the output signal?
10. Turn the potentiometer fully clockwise. What happens to the output signal?
11. Turn the potentiometer until the output signal on the collector shows no signs of distortion.
12. Record the output voltage on Table 4-8-1.
13. Connect the oscilloscope probe to the base of the transistor.
14. Record the input voltage on Table 4-8-1.
15. Calculate the voltage gain of the circuit using the following formula:

$$X_C = \frac{E_{OUT}}{E_{IN}}$$

16. Record the voltage gain on Table 4-8-1.

Directions

Use Figure 4-8-1 for steps 1 - 16 on page 149.

FIGURE 4-8-1

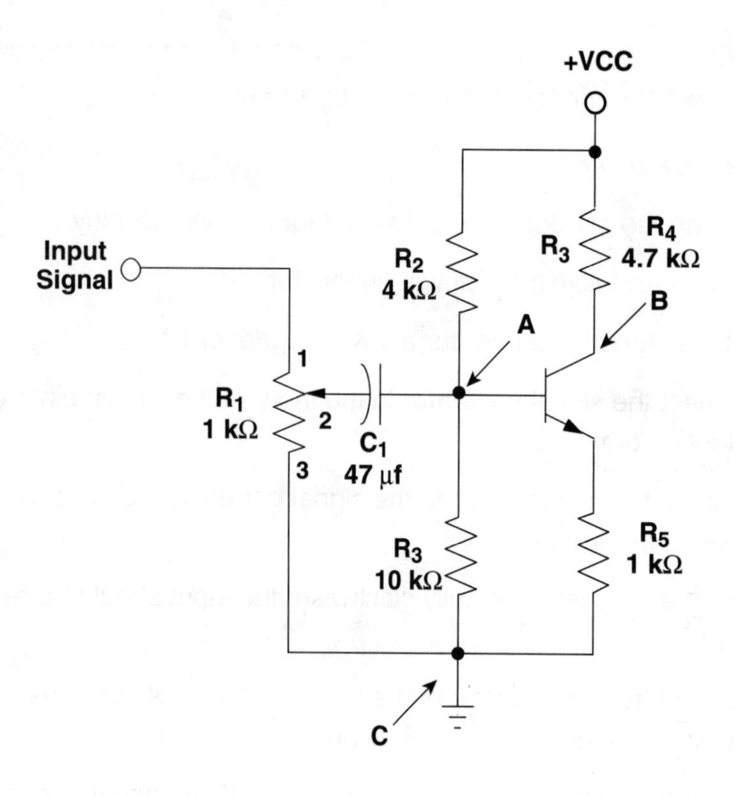

TABLE 4-8-1

	E_{OUT}	E_{IN}	E_{GAIN}
1			
2			
3			

TABLE 4-8-2

Step	What happens to the output signal?
9	
10	

LAB 4-8

Common Emitter Amplifier

Summary

The circuit constructed was a common emitter amplifier using a voltage divider bias with emitter feedback. A signal was applied to the input of the amplifier and the amplitude was increased while observing the output on the oscilloscope. When the input signal was increased to the maximum value, the positive and negative alternations were flattened out. This occurred because the amplifier was being overdriven. The output signal was then adjusted to a point where no distortion occurred and the output and input voltage were recorded. The voltage gain should have been approximately 4.5.

LAB 4-9

Oscillators

Name
Course
Date Due

Objective

The student should be able to identify the terminology and circuits associated with oscillators.

Reference

Chapter 30, pages 282-300

Materials Required

None

Equipment Required

None

Notes

Definitions

Define the following terms in complete sentences.

1. Oscillator

2. Tank Circuit

3. Positive Feedback

4. Sinusoidal Oscillator

5. Crystal

6. Nonsinusoidal Oscillator

7. Relaxation Oscillator

8. Multivibrator

9. Astable Multivibrator

Questions

Answer the following questions in complete sentences.

1. What are three types of LC oscillators?

2. What are two types of RC oscillators?

3. What integrated circuit can be readily used as a nonsinusoidal oscillator with very few discrete components?

4. Draw a block diagram and label all parts of a basic oscillator.

LAB 4-9

Oscillators

Notes

LAB 4-10

555 Timer

Name

Course

Date Due

Objective

The student should be able to determine the operation and characteristics of a 555 timer integrated circuit.

Reference

Chapter 30, pages 287-289

Materials Required

555 Timer IC
820 Ω, 1/2 W Resistor
150 Ω, 1/2 W Resistor
10 kΩ, 1/2 W Resistor
470 kΩ, 2 W Potentiometer
1 μF Electrolytic Capacitor

Equipment Required

Perf Board
Oscilloscope
  with Test Probe
Power Supply

Notes

Procedure

As each step is completed, check it off.

1. Connect the circuit shown in Figure 4-10-1.
2. Turn on the oscilloscope and adjust for a proper display.
3. Turn on and adjust the power supply for +10 V.
4. Connect the oscilloscope to pin 3 of the 555 timer.
5. Draw the waveform on Figure 4-10-2.
6. Record the voltage and the time to complete one cycle on Table 4-10-1.
7. Determine the frequency of the circuit using the following formula:

$$f = \frac{1}{t}$$

8. Change the values of R_1 and R_2 to the value indicated on Table 4-10-1.
9. Repeat steps 3 - 9 for each set of values listed on Table 4-10-1.

LAB 4-10

555 Timer

Directions

Complete Figure 4-10-2 and Table 4-10-1 using steps 1 - 9 on page 153.

FIGURE 4-10-1

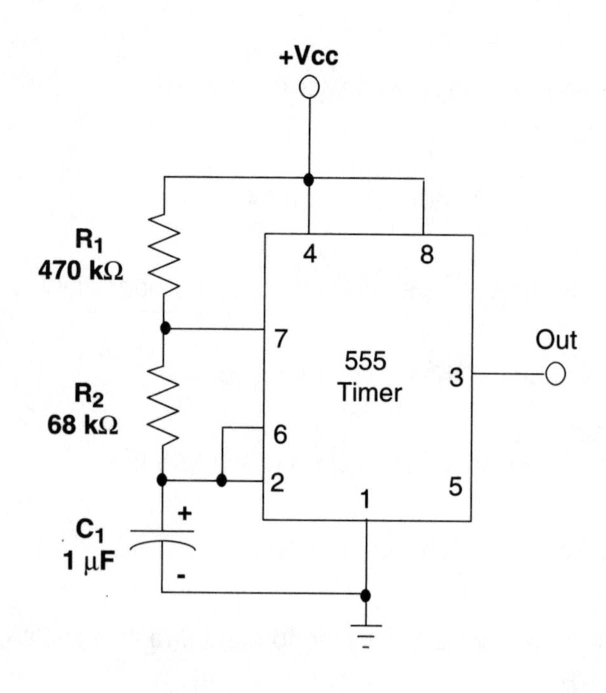

FIGURE 4-10-2

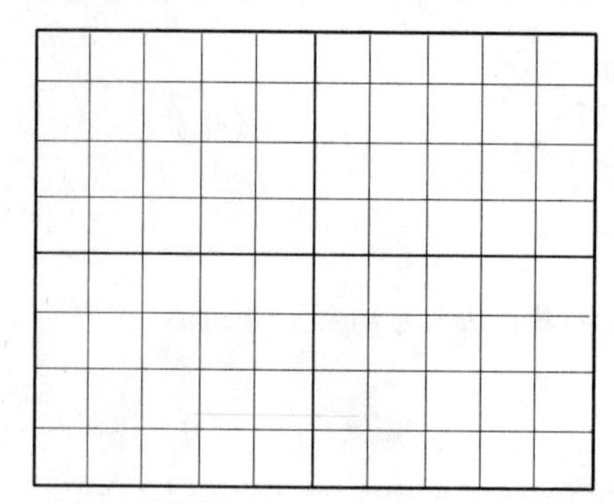

Scope Setting

Volts/cm	
Time/cm	

Probe Setting

Measured Voltage	
Across What Points	

TABLE 4-10-1

R_1	R_1	E_{OUT}	t	f
470 kΩ	68 kΩ			
10 kΩ	1500 Ω			
820 Ω	150 Ω			

Summary

The 555 timer was connected to astable operation. The timing is split between resistor R_1 and R_2 with pin 7 connected to the junction of R_1 and R_2. When the power was applied, capacitor C_1 charged through resistors R_1 and R_2 and discharged through R_2 until a new cycle was started. By making R_2 larger than R_1, a symmetrical square wave can be obtained with a duty cycle of 50%.

LAB 4-11

Waveshaping Circuits

Name
Course
Date Due

Objective

The student should be able to identify the terminology and circuits associated with waveshaping circuits.

Reference

Chapter 31, pages 292-300

Materials Required

None

Equipment Required

None

Notes

Definitions

Define the following terms in complete sentences.

1. Time Domain
2. Frequency Domain
3. Harmonics
4. First Harmonics
5. Periodic Waveforms
6. Pulse Width
7. Duty Cycle
8. Overshoot
9. Undershoot
10. Ringing
11. Differentiator
12. Integrator
13. Clipping Circuit
14. Clamping Circuit

Definitions

Define the following terms in complete sentences.

15. Monostable Multivibrator

16. Bistable Multivibrator

17. Schmitt Trigger

Questions

Answer the following questions in complete sentences.

1. Draw a schematic diagram of a differentiator.

2. Draw a schematic diagram of an integrator.

3. Draw a schematic diagram of a clipping circuit.

4. Draw a schematic diagram of a clamping circuit.

Notes

LAB 5-1

Binary Number System

Name

Course

Date Due

Objective

The student should be able to identify and use the terminology and characteristics associated with the binary number system.

Reference

Chapter 32, pages 303-308

Materials Required

None

Equipment Required

None

Notes

Definitions

Define the following terms in complete sentences.

1. Binary Number System

2. Bit

3. LSB

4. MSB

5. Binary Coded Decimal

LAB 5-1

Binary Number System

Notes

LAB 5-2

Binary & Decimal Conversion

Name

Course

Date Due

Objective

The student should be able to accurately convert numbers between the binary and decimal number systems.

Reference

Chapter 32, pages 305-306

Materials Required

None

Equipment Required

None

Directions

Convert the following binary numbers to decimal numbers.

1. 10011011

2. 11110010

3. 01010101

4. 10001110

5. 10001111 01100110

6. 11110001 01111011

Directions

Convert the following decimal numbers to binary numbers.

1. 476

2. 1056

3. 12,789

4. 390,472

5. 35.78

6. 128.128

LAB 5-2

Binary & Decimal Conversion

Notes

LAB 5-3

Basic Logic Gates

Name

Course

Date Due

Objective

The student should be able to identify and use the terminology and characteristics associated with basic logic gates.

Reference

Chapter 33, pages 309-315

Materials Required

None

Equipment Required

None

Notes

Definitions

Define the following terms in complete sentences.

1. Truth Table

2. NOT Gate

3. AND Gate

4. OR Gate

5. NAND Gate

6. NOR Gate

7. Exclusive OR Gate

8. Exclusive NOR Gate

Questions

Draw the required symbol and truth tables.

1. Draw the AND gate symbol and develop the truth table for it.

2. Draw the OR gate symbol and develop the truth table for it.

3. Draw the NAND gate symbol and develop the truth table for it.

4. Draw the NOR gate symbol and develop the truth table for it.

5. Draw the XNOR gate symbol and develop the truth table for it.

Notes

LAB 5-4

Basic Logic Circuits

Name

Course

Date Due

Objective

The student should be able to connect and develop truth tables for basic logic circuits.

Reference

Chapter 33, pages 309-315

Materials Required

2, SPDT Switches

6 V Relay

Lamp and Socket

Equipment Required

6 V Power Supply

Notes

Procedure

As each step is completed, check it off.

1. Connect the circuit shown in Figure 5-4-1. Apply power to the circuit.
2. Using Table 5-4-1, set the switches to the positions indicated and record the results.
3. Remove power and disconnect.
4. Connect the circuit shown in Figure 5-4-2. Apply power to the circuit.
5. Using Table 5-4-2, set the switches to the positions indicated and record the results.
6. Remove power and disconnect the circuit.
7. Connect the circuit shown in Figure 5-4-3. Apply power to the circuit.
8. Using Table 5-4-3, set the switches to the positions indicated and record the results.
9. Remove power and disconnect the circuit.
10. Connect the circuit shown in Figure 5-4-4. Apply power to the circuit.
11. Using Table 5-4-4, set the switches to the positions indicated and record the results.
12. Remove power and disconnect circuit.
13. Connect the circuit shown in Figure 5-4-5. Apply power to the circuit.
14. Using Table 5-4-5, set the switches to the positions indicated and record the results.
15. Remove power and disconnect the circuit.
16. Connect the circuit shown in Figure 5-4-6. Apply power to the circuit.
17. Using Table 5-4-6, set the switches to the positions indicated and record the results.
18. Remove power and disconnect circuit.

Directions

Use Figure 5-4-1 and Table 5-4-1 for steps 1 - 3 on page 163, and Figure 5-4-2 and Table 5-4-2 for steps 4 - 6 on page 163.

FIGURE 5-4-1

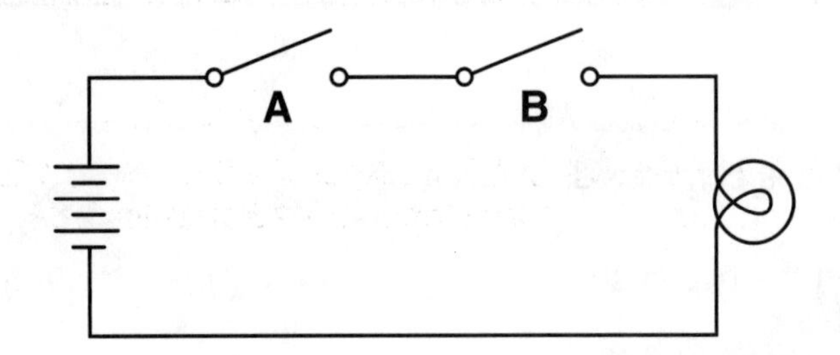

TABLE 5-4-1

Switch Position		Condition of Light
A	B	OFF (0) or ON (1)
Open	Open	
Close	Open	
Open	Close	
Close	Close	

FIGURE 5-4-2

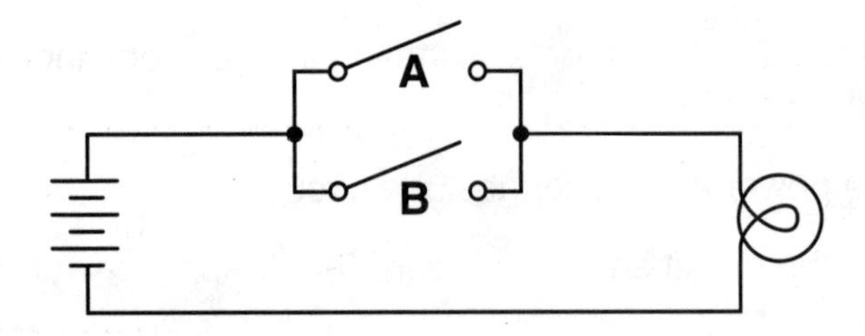

TABLE 5-4-2

Switch Position		Condition of Light
A	B	OFF (0) or ON (1)
Open	Open	
Close	Open	
Open	Close	
Close	Close	

LAB 5-4

Basic Logic Circuits

Notes

LAB 5-4

Basic Logic Circuits

Notes

Name	
Course	
Date Due	

Directions

Use Figure 5-4-3 and Table 5-4-3 for steps 7 - 9 on page 163, and Figure 5-4-4 and Table 5-4-4 for steps 10 - 12 on page 163.

FIGURE 5-4-3

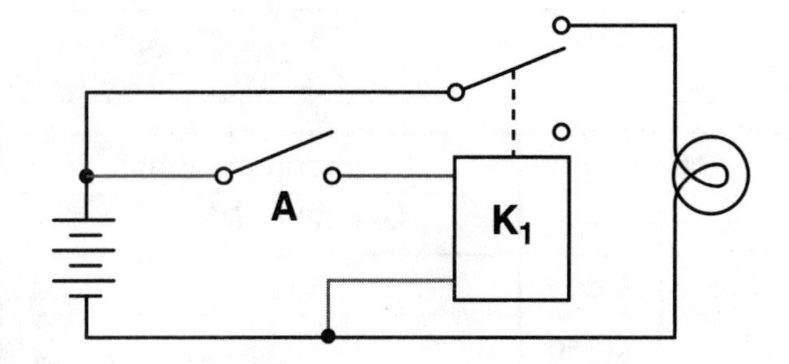

TABLE 5-4-3

Switch Position	Condition of Light
A	OFF (0) or ON (1)
Open	
Close	

FIGURE 5-4-4

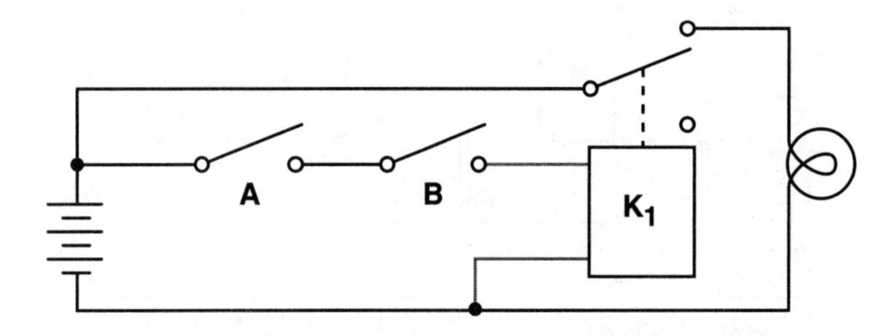

TABLE 5-4-4

Switch Position		Condition of Light
A	B	OFF (0) or ON (1)
Open	Open	
Close	Open	
Open	Close	
Close	Close	

Directions

Use Figure 5-4-5 and Table 5-4-5 for steps 13 - 15 on page 163, and Figure 5-4-6 and Table 5-4-6 for steps 16 - 18 on page 163.

FIGURE 5-4-5

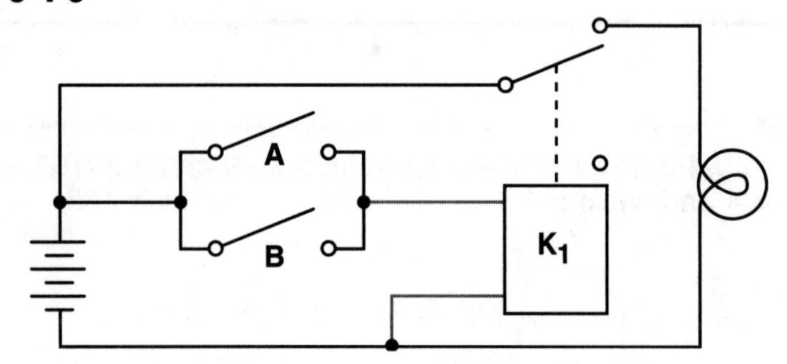

TABLE 5-4-5

Switch Position		Condition of Light
A	B	OFF (0) or ON (1)
Open	Open	
Close	Open	
Open	Close	
Close	Close	

FIGURE 5-4-6

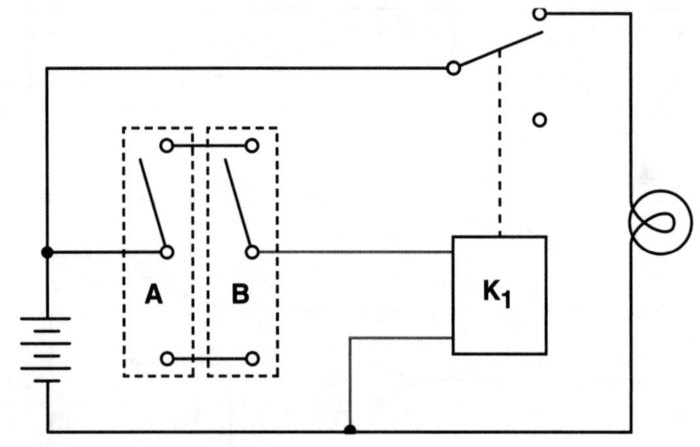

TABLE 5-4-6

Switch Position		Condition of Light
A	B	OFF (0) or ON (1)
Open	Open	
Close	Open	
Open	Close	
Close	Close	

LAB 5-4

Basic Logic Circuits

Summary

The circuit connected in steps 1 - 3 was an AND gate. The AND gate produces an output only when all its inputs are 1's. If any of the inputs are 0's, the output is 0. When the light is out, no current is flowing and the output is a 0. When the light is lit, current is flowing and the output is a 1.

The circuit connected in steps 4-6 was an OR gate. The OR gate produces a 1 output if any of its inputs are 1's. The output is a zero if all the inputs are 0's.

The circuit connected in steps 7 - 9 was a NOT gate or inverter. The NOT performs the function of inversion, or complementation. The purpose of the inverter is to make the output state the opposite of the input state.

The circuit connected in steps 10 - 12 was a NAND gate. The NAND gate is a combination of an inverter and an AND gate. The NAND gate is the most commonly used logic function. Notice that the output of the NAND gate is the complement of the output of the AND gate.

The circuit connected in steps 13 - 15 was a NOR gate. The NOR gate is a combination of an inverter and an OR gate. Notice that the output of the NOR gate is the complement of the OR function output.

The circuit connected in steps 16 - 18 was an XOR gate or exclusive OR gate. The XOR gate has only two inputs. It differs from the OR gate in that it produces a 3 output if both inputs are 0's or 1's. It generates a 1 output if either but not both inputs is a 1.

LAB 5-5A

Simplifying Logic Circuits

Name

Course

Date Due

Objective

The student should be able to use Veitch diagrams in simplifying logic circuits.

Reference

Chapter 34, pages 316-319

Materials Required

None

Equipment Required

None

Notes

Directions

Use the following steps in simplifying the following Boolean expression using Veitch diagrams.

1. Draw the diagram based on the number of variables.

2. Plot the logic functions by placing an X in each square representing a term.

3. Obtain the simplified logic functions by looping adjacent groups of X's in groups of eight, four or two. Continue to loop until all X's are included in a loop.

4. "OR" the loops with one term per loop. (Each expression is pulled off the Veitch diagram and "OR" ed using the "+" symbol, i.e. ABC + DEF).

5. Write the simplified expression.

Directions

Use the above steps to help solve the following questions.

1. Reduce $A\overline{B} + \overline{A}B + AB$ to its simplest form.

Directions

Use the above steps to help solve the following questions.

2. Reduce $AB\overline{C} + ABC + A\overline{B}C + \overline{A}BC$ to its simplest form.

3. Reduce the following to its simplest form:
$ABC\overline{D} + AB\overline{C}D + ABCD + \overline{A}BCD + A\overline{B}D + A\overline{B}\,\overline{C}\,\overline{D} + \overline{A}\,\overline{B}\,\overline{C}\,\overline{D}$

LAB 5-5A

Simplifying Logic Circuits

Notes

LAB 5-5B

Simplifying Logic Circuits

Name

Course

Date Due

Objective

The student should be able to use Karnaugh maps in simplifying logic circuits.

Reference

Chapter 34, pages 319-323

Materials Required

None

Equipment Required

None

Notes

Directions

Use the following steps in simplifying the following Boolean expression using Veitch diagrams.

1. Draw the diagram based on the number of variables.

2. Plot the logic functions by placing an X in each square representing a term.

3. Obtain the simplified logic functions by looping adjacent groups of X's in groups of eight, four, or two. Continue to loop until all X's are included in a loop.

4. "OR" the loops with one term per loop. (Each expression is pulled off the Karnaugh map and "OR" ed using the "+" symbol, i.e., ABC + DEF).

5. Write the simplified expression.

Directions

Use the above steps to help solve the following questions.

1. Reduce $A\overline{B} + \overline{A}B + AB$ to its simplest form.

Directions

Use the above steps to help solve the following questions.

2. Reduce $AB\overline{C} + ABC + A\overline{B}C + \overline{A}BC$ to its simplest form.

3. Reduce the following to its simplest form:

   $ABC\overline{D} + AB\overline{C}D + ABCD + \overline{A}BCD + A\overline{B}D + A\overline{B}\,\overline{C}\,\overline{D} + \overline{A}\,\overline{B}\,\overline{C}\,\overline{D}$

LAB 5-5B

Simplifying Logic Circuits

Notes

LAB 5-6

Sequential Logic Circuits

Name
Course
Date Due

Objective
The student should be able to identify and use the terminology and characteristics associated with sequential logic circuits.

Reference
Chapter 35, pages 325-340

Materials Required
None

Equipment Required
None

Notes

Definitions
Define the following terms in complete sentences.

1. Flip-Flop
2. RS Flip-Flop
3. Clocked Flip-Flop
4. D Flip-Flop
5. JK Flip-Flop
6. Latch
7. Counter
8. Modulus
9. Asynchronous
10. Ripple Counter
11. Synchronous
12. Parallel Counter

Definitions

Define the following terms in complete sentences.

13. Decade Counter

14. Up-down Counter

15. Shift Register

Questions

Draw the required symbols and truth tables.

1. Draw the RS Flip-flop logic symbol and develop a truth table for it.

2. Draw the D Flip-flop logic symbol.

3. Draw the JK Flip-flop logic symbol.

4. Draw the logic symbol for a four-stage counter.

Notes

LAB 5-7

Combinational Logic Circuits

Name

Course

Date Due

Objective

The student should be able to identify the terminology and circuits associated with combinational logic circuits.

Reference

Chapter 36, pages 341-355

Materials Required

None

Equipment Required

None

Notes

Definitions

Define the following terms in complete sentences.

1. Encoder
2. Priority Encoder
3. Decoder
4. Multiplexer
5. Adder
6. Half Adder
7. Full Adder
8. Subtractor
9. Half Subtractor
10. Full Subtractor
11. Comparator

Questions

Draw the required symbols.

1. Draw the logic symbol for a decimal-to-binary priority encoder.

2. Draw the logic symbol for a binary-to-decimal decoder.

3. Draw the logic symbol for a sixteen-input multiplexer.

4. Draw the logic symbols for a half and full adder.

5. Draw the logic symbol for a half and full subtractor.

6. Draw the logic symbol for a four-bit comparator.

Notes

LAB 5-8

Microcomputer Basics

Name

Course

Date Due

Objective

The student should be able to identify the terminology and circuits associated with microcomputer basics.

Reference

Chapter 37, pages 357-365

Materials Required

None

Equipment Required

None

Notes

Definitions

Define the following terms in complete sentences.

1. Digital Computer
2. Control
3. Arithmetic Logic Unit
4. Memory
5. Input/Output
6. Program
7. RAM
8. ROM
9. Microprocessor
10. Accumulator
11. Condition Code Register

Definitions

Define the following terms in complete sentences.

12. Program Counter

13. Stack Pointer

14. Instructions

Questions

Draw the required diagrams.

1. Draw a block diagram of a digital computer.

2. Draw a block diagram of the ALU (Arithmetic Logic Unit).

3. Draw a block diagram of an 8-bit microprocessor.

Notes

LAB 6-1

Soldering Review

Name

Course

Date Due

Notes

Review

In electronics, the basic goal of soldering is to electrically and mechanically join two circuit components. For soldering to be successful and reliable, the solder must adhere to the mating surface. The application of solder to a base metal requires the surface to be clean and free of contamination. If the surface to be soldered is contaminated, the solder tends to ball up and not adhere. Principal sources of contamination are greases, oils, and dirt that prevent good soldering adhesion. Aging also causes surface contamination by the formation of an oxide film.

Technically, soldering is the joining of two pieces of metal alloy having a melting point below 800 degrees Fahrenheit. Solder includes a combination of tin and lead. Tin-lead solder ranges from pure tin to pure lead and all proportions in between. In electrical soldering, the alloy mix is usually 60% tin and 40% lead (60/40).

Characteristics of alloys of tin and lead are plotted against temperature in Figure 6-1-1. This graph allows one to see that only an alloy of 63/37 has a eutectic point (a single melting point). All other combinations start melting at one temperature, pass through a plastic stage, and then become a liquid at a higher temperature. Any physical movement of the components being soldered while the solder is in the plastic stage will result in a cold solder connection. Such a connection will appear dull and grainy, and is mechanically weaker and less reliable. Therefore, 63/37 or 60/40 solder is commonly used in electronics because it does not remain in the plastic stage very long.

FIGURE 6-1-1

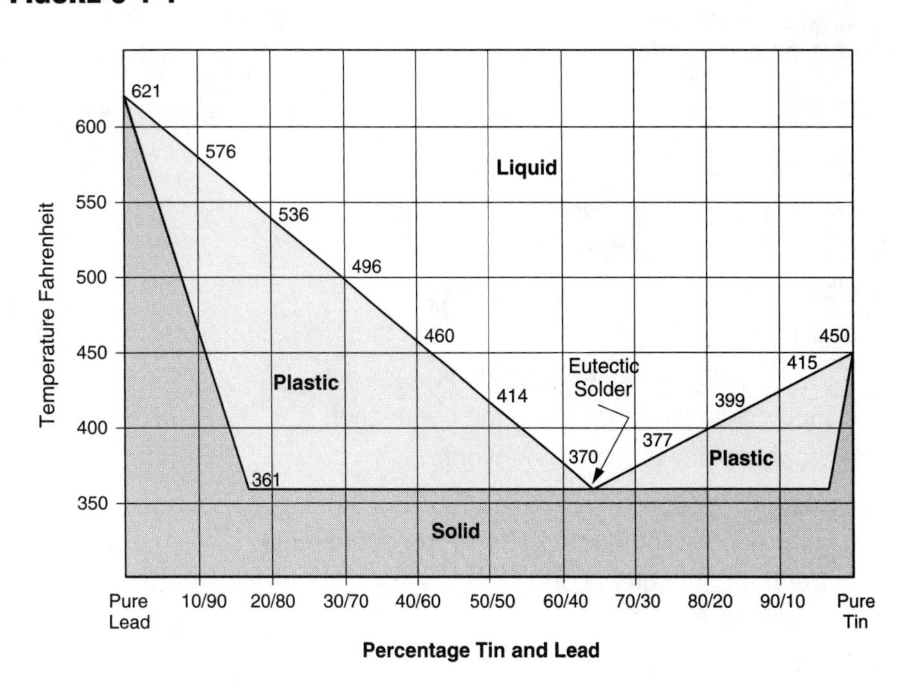

Review (continued)

Flux helps the solder alloy flow around the connections. Flux cleans the component leads of oxide and film allowing the solder to adhere. Soldering flux is usually included in the solder in the central core. Rosin flux is always used in electrical soldering because it is noncorrosive.

The soldering iron is the primary tool used in soldering. The soldering iron is often called a pencil iron because it resembles a thick pencil. The soldering iron is usually specified by wattage. In most cases a 25 to 35 W soldering iron is adequate. Wattage represents the amount of heat capacity available at the tip. Soldering irons of all wattages usually run at the same temperature. A low-wattage soldering iron will tend to cool faster during the soldering process than a high-wattage soldering iron. Soldering irons can produce static voltage spikes that will destroy many integrated circuit components. A grounded soldering iron tip is a wise safety measure.

Soldering iron tips are usually selected by preference. Each type and shape has its place and purpose. The commonly used chisel tip is a versatile and convenient tip.

The best soldering techniques can be outlined simply. First, wipe the tip of the soldering iron on a damp sponge. Then place a small amount of solder on the tip and apply the tip to the connection to be soldered. Solder should not be applied until the connection becomes hot enough to melt the solder. How long this takes is quickly learned by a few trials. When the solder has melted and flowed into a contoured fillet, remove the solder. Keep the tip of the soldering iron on the connection for a few seconds longer, then remove it. Do not disturb the newly made connection until it has had time to solidify. A good solder connection will be bright and shiny. Disturbing the connection before the solder has had time to solidify will produce a cold solder connection.

Because your fingers may be dirty and oily, handle components and the printed circuit board as little as possible. If there is a question about the cleanliness of a part, clean it using alcohol, steel wool, or scouring powder. When using scouring powder, ensure that the parts are rinsed thoroughly. If steel wool is used, use a lint-free cloth to remove all pieces of the wool from the part or printed circuit board.

Summary

1. Use clean parts and a clean printed circuit board.
2. Use 60/40 or 63/37 rosin core solder.
3. Use a 25- to 35-watt soldering iron.
4. Use the proper soldering sequence.
    a. Clean soldering iron tip.
    b. Apply solder to soldering iron tip.
    c. Apply soldering iron tip to the connection.
    d. Apply solder to the connection.
    e. Remove the solder.
    f. Remove soldering iron tip from the connection.
    g. Let the connection cool.
5. Use patience.
6. Practice.

Notes

PROJECT 1

Electronic Siren

Name

Course

Date Due

Notes

Overview

What is an electronic siren?

An electronic siren is an electronic circuit that can simulate many types of siren sounds. The more common siren sounds that can be simulated are the electronic siren, the warble, the two-tone, the pulse, and the yowl. Numerous other sounds can also be produced by the electronic siren.

General Information

The electronic siren consists of three basic blocks. Two of the blocks are astable multivibrators (IC_1 and IC_2) and the third block is a power amplifier (Q_1) as shown in Figure 1. The astable multivibrators, IC_1 and IC_2, are 555 integrated circuit timers.

FIGURE 1

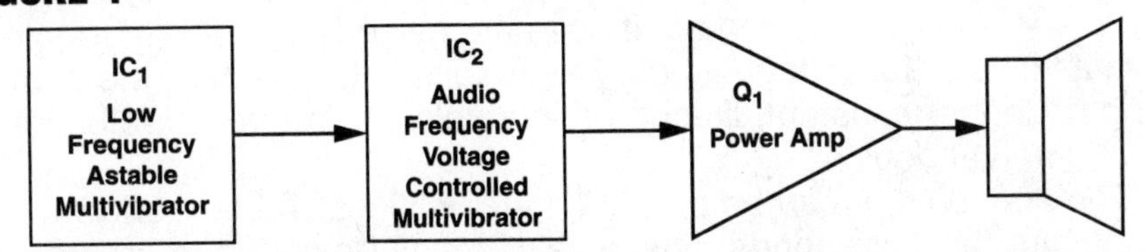

An astable multivibrator is a free-running device that does not require a trigger to start it. The output is a square wave. The time the output goes from high to low is determined by resistor R_1 and capacitor C_1, Figure 2.

FIGURE 2

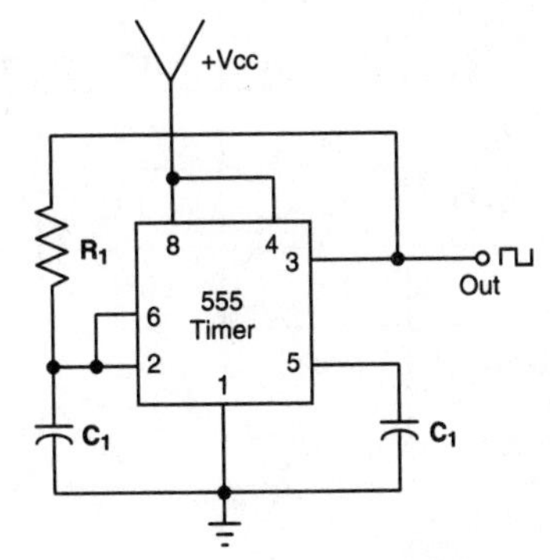

The timing period is determined by: $t = 1.4(R1)(C1)$

The output frequency can be determined by: $f = \frac{1}{t}$

In the schematic, timer IC_1 is connected as a low-frequency astable multivibrator. In this configuration the square wave output at pin 3 remains fixed at approximately 50 percent of the duty cycle as its frequency is varied by timing potentiometer R_1. With the exception of the two-tone and pulse mode the low-frequency square wave output is integrated by R-C integrator potentiometer R_2 and capacitors C_5, C_6, C_7, or C_8. This produces a triangular wave at pin 5 of IC_2. Timer IC_2 is configured as an audio-frequency voltage-controlled multivibrator.

General Information (continued)

Its frequency is controlled by the voltage on pin 5 (control voltage input) and by timing resistor R_3. The triangular waveform at pin 5 frequency modulates IC_2 so that the square wave output at pin 3 rises and falls in frequency to duplicate the familiar wail of a siren.

The amplifier Q_1 increases the output from pin 3 of IC_2. Capacitor C_9 and resistor R_4 form a low-pass filter to remove some of the high-frequency components from the square wave output.

The *two-tone* mode uses a square wave output from IC_1. The square wave is used to frequency modulate IC_2. Potentiometer R_2 is used to attenuate (reduce) the square wave causing IC_2 to shift frequency abruptly at a rate determined by IC_1. The resulting sound created is a distinctive "twee-dee" similar to a European police siren.

The *pulse mode* also uses IC_1's square wave output, but not to frequency modulate IC_2. In this mode the square wave is routed to pin 4 (reset) of IC_2. As long as pin 4 is held high, IC_2 will stop running. The square wave will gate IC_2 on and off at the frequency of IC_1. The pitch of the sound will be constant because the frequency of IC_2 is determined only by the value of timing resistor R_3.

The *yowl mode* is a combination of the siren and pulse modes. The square wave from IC_1 is routed to the reset pin of IC_2. This signal is also integrated by capacitor C_6 and potentiometer R_2 at pin 5 of IC_2. IC_2 is gated on and off as in the pulse mode. The difference is that every time IC_2 is gated on, it sees the rising half of the triangular wave as its control voltage input. This results in the pitch of the sound being no longer constant. During on intervals the frequency falls until the off interval begins.

The sounds that can be produced are limitless. Table 1 shows settings for familiar sounds. These values may be varied to create even more sounds. In all cases, final tweaking is necessary to produce the desired sound.

TABLE 1

Mode	R_1 Rate	R_2 Range	R_3 Pitch	S_3 Integrator	S_2
Electronics Siren	470kΩ	10kΩ	120kΩ	100μfd	Off
Warble	100kΩ	100kΩ	100kΩ	10μfd	Off
Two-Tone	560kΩ	10kΩ	150kΩ	NA	Off
Pulse	680kΩ	1kΩ	100kΩ	NA	On
Yowl	1MΩ	1kΩ	100kΩ	500μfd	On

The rate at which the sound varies is determined by the value of IC_1's timing potentiometer R_1. Increasing the value of R_1 will decrease the rate and vice versa. The pitch of the output is determined by both the amplitude of the modulating control voltage and the value of IC_2 timing potentiometer R_3. As the control voltage increases in amplitude and/or the value of R_3 is increased, the pitch decreases and vice versa.

Notes

PROJECT 1

Electronic Siren

Name

Course

Date Due

Notes

General Information (continued)

The range of modulation, and the difference between the high and low frequency of IC_2 is set by the value of potentiometer R_2. A small value of R_2 permits a large range, while a large value restricts it. About 1000 Ω is the practical minimum value of R_2, as well as R_1 and R_3.

The range is also controlled by the size of the integrating capacitor C_5, C_6, C_7, and C_8. The product of the capacitor value and potentiometer R_2 is the time constant of the integrator. The time constant establishes the linearity of each half of the triangular wave can rise to during the on-time of IC_1. Because this is frequency-dependent, there will be some interaction among potentiometers R_1 and R_2 and capacitors C_5, C_6, C_7, and C_8 as the range is set.

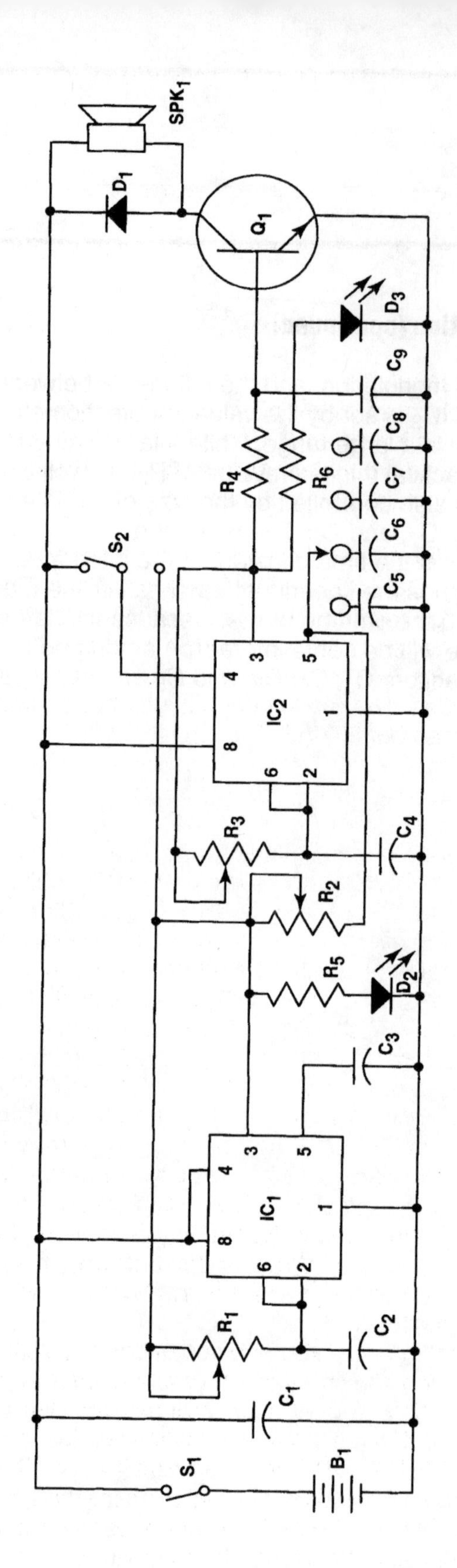

PROJECT 1

Electronic Siren

Parts List

Resistors:

- R_1, 1 MΩ Potentiometer
- R_2, 100 kΩ Potentiometer
- R_3, 250 kΩ Potentiometer
- R_4, 220 Ω, 1/2 W Resistor
- R_5, R6, 560 Ω ,1/2 W Resistor
- R_7, 100 Ω 1/2 W Resistor

Capacitors:

- C_1, C5, 10 µF, 50 V Electrolytic
- C_2, C9, 1 µF, 50 V Electrolytic
- C_3, C4, 0.01 µF, 50 V Ceramic Disk
- C_6, 100 µF, 50 V Electrolytic
- C_7, 500 µF, 6 V Electrolytic
- C_8, 1000 µF, 25 V Electrolytic

Semiconductors:

- D_1, 1N4003 or Equivalent
- D_2, D_3, Jumbo Red LED
- IC_1, IC_2, 555 IC Timer
- Q_1 , TIP31 Power Transistor

Miscellaneous:

- S_1, DPDT Switch
- S_2, SPDT Switch
- S_3, 8-Position Rotary Switch
- SPK_1, 8 Ω, 3-inch Speaker

PROJECT 2

Power Supply

Name
Course
Date Due

Notes

Overview

What is a power supply?

A power supply is an electronic circuit that converts 120 V AC to a DC voltage required for a particular application. The voltages are regulated to maintain a constant output voltage. The circuit design allows for a fixed voltage output as well as a variable voltage output.

General Information

Diodes D_1, D_2, D_3, and D_4 form a bridge rectifier. The diode bridge BR_1 is also a bridge rectifier. The resulting ripple frequency in both the diode bridge rectifier and the diode bridge is 120 Hz.

The diode bridge BR_1 is easier to install on the circuit board than the individual diodes. The diode bridge is purchased according to the highest PIV that can be applied. The smallest rating is 50 V. The current handling capabilities physically determine the size of the package. The smallest current handling size is 112 A. The unit that was installed is rated at 50 V, 1 A.

Filter capacitors C_1 and C_5 have large values of capacitance. The large values of the filter capacitors create a very small ripple frequency. Capacitor C_5 is twice as large as capacitor C_1. Therefore, the output from capacitor C_5 is a much smaller ripple frequency with a slightly higher output voltage for the same applied input voltage.

The LM309K is a fixed +5 V voltage regulator. The LM309 is available in two packages, K or H. It features internal current limiting and thermal shutdown. Capacitor C_6 is required if the regulator is located more than four inches from the power supply filter. It is a good idea to include capacitor C_6 even if the filter and regulator are close together. Capacitors C_7 and C_8 are used to improve the transient response. Diode D_{11} and D_{12} are used only for protection of the regulator. Resistor R_3 and LED_2 give a visual indication that the power supply is working. When the power supply is shut off, the resistor and LED provide a path for the current to bleed down. This is important. There should be no voltage present at the terminals when hooking the power supply to a circuit.

The LM317 is an adjustable 9-terminal positive voltage regulator capable of supplying in excess of 1.5 amperes over an output voltage range of 1.2 V to 37 V. Only two external resistors are required to set the voltage. Figure 4 is a simple schematic that illustrates the operation of the LM317. The integrated circuit keeps the voltage drop between the output terminal and the adjustment terminal at a constant 1.25 V. Resistor R_1 is connected between these two terminals, setting up a constant adjustment current. The magnitude of this adjustment current and the setting of potentiometer R_2 determine the output voltage of the regulator. Any voltage greater than 1.25 V can be obtained by increasing the resistance between the adjustment terminal and ground.

General Information (continued)

Capacitor C_2 is only required if the voltage regulator is located more than four inches from the filter capacitor. It is a good idea to include C_2 even if the filter capacitor is close to the regulator. Resistor R_1 establishes an adjustment current of 16 mA which flows through the voltage adjustment potentiometer R_2. Adjusting R_2 for maximum resistance places the adjustment terminal at 18 V above ground. This sets the power supply output voltage at 19.25 V. Capacitor C_3 filters out any ripple voltage at the adjustment terminal. Transient response and stability are improved by the addition of capacitor C_4. Diode D_6 provides a discharge path for capacitor C_3 in the event of a short circuit across the output. Diode D_5 and D_7 protect the IC regulator against any reverse voltage that might accidentally be applied at the output.

As the resistance connected to the power supply decreases, the current increases when the voltage remains the same. This supports Ohm's Law, I=E/R. Also, as the current increased, the amount of power dissipated as heat increased. This was supported by the use of Watt's Law, P=IE.

When using a power supply, it is important to watch the amount of current demanded by the load. If the current is too large, the internal components of the power supply will dissipate more heat. If too much heat is dissipated, the IC will shut off through the use of a device called thermal overload which is built internally in the IC.

When using a power supply it is a good practice, when hooking up an unknown load, to use an ammeter and voltmeter to monitor the current and voltage demands. This not only protects the load from damage, it also protects the power supply.

This power supply is good for a fixed 5 V at 1 A from the LM309K. If a load is connected that draws more than one ampere, remove it and inspect it for problems. If the load is okay, then use a power supply that will supply more than one ampere.

Notes

PROJECT 2

Power Supply

Name

Course

Date Due

Parts List

Resistors:

- R_1, 2 kΩ, 2 W Potentiometer
- R_2, 150 Ω, 1/2 W Resistors
- R_3, R4, 1 kΩ, 1/2 W Resistors

Capacitors:

- C_1, 1000 µF, 25 V Electrolytic
- C_2, C_4, C_6, C_8, 0.1 µF, 50 V Ceramic Disk
- C_3, C_7, 10 µF, 50 V Electrolytic
- C_5, 2200 µF, 25 V Electrolytic

Semiconductors:

- D_1 - D_9, 1N4003 Silicon Diode
- BR_1, 50 V, 1 A Diode Bridge Rectifier
- IC_1, LM317 Variable Voltage Regulator
- IC_2, LM309K Fixed 5 V Voltage Regulator
- LED_1, LED_2, Jumbo Red LED

Miscellaneous:

- S1, SPST Switch
- S2, 4PDT Switch
- 115 V/18 V/6 V Transformer
- 1/4 A Fuse
- Fuse Holder
- 1 Red Banana Jack
- 1 Black Banana Jack
- Power Cord with Plug
- T03 Heat Sink
- T220 Heat Sink

Schematic

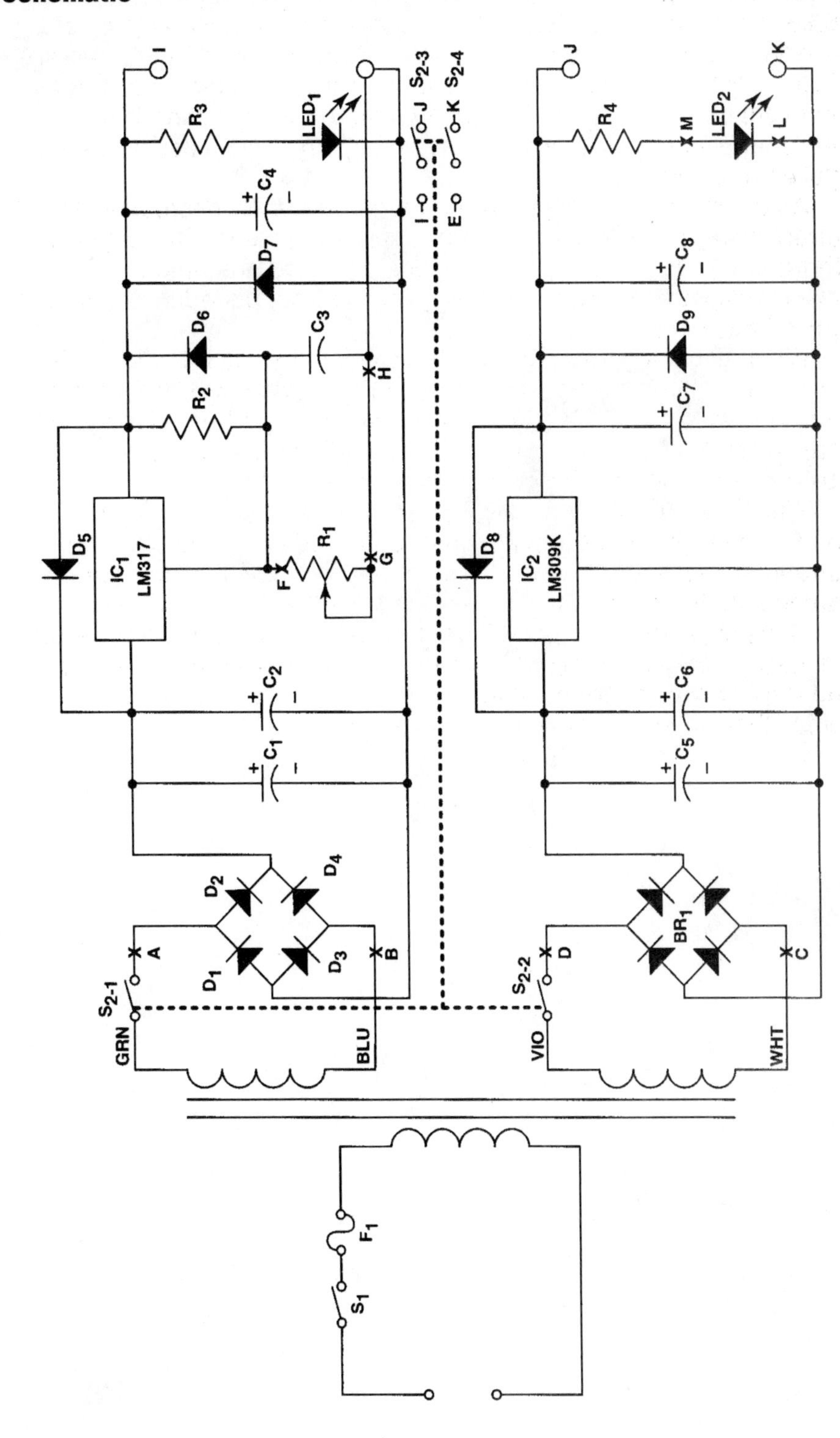

Notes

PROJECT 3

Electronic Color Organ

Notes

Name

Course

Date Due

Overview

What is a color organ?

A color organ control is an electronic circuit that allows the audio signal from a stereo amplifier to drive three separate lamp circuits. The three separate lamp circuits are controlled by the low, medium, and high frequencies of the music from the stereo amplifier.

General Information

The purpose of this project is to apply the skills and knowledge acquired to date. The project covers the following areas: printed circuit board fabrication, soldering, component identification, use of power and hand tools, screen printing, use of the oscilloscope, use of the function generator, use of the isolation transformer, use of the multimeter, circuit analysis, and troubleshooting.

***WARNING**: An isolation transformer must be used while making all measurements. You will be working with 110 V AC. A serious shock potential does exist. Be sure the circuit is unplugged before making any adjustments.*

The color organ control receives an audio input signal from an audio amplifier. The audio signal is applied to transformer T_1. The transformer isolates the amplifier from the color organ control and provides the first stage of amplification. The amplified audio signal is applied to a voltage divider that consists of resistor R_1 and potentiometer V_{R_1}. The signal is then capacitively coupled to the audio amplifier, transistor Q_1. Resistors R_2 and R_3 establish the bias points for transistor Q_1. Resistor R_4 acts as a load and current-limiting resistor for transistor Q_1. The output of the audio amplifier is amplified over what was at the input.

The power supply provides the necessary voltage for operation of the audio amplifier. It consists of diode D_1, voltage dividing resistors R_5 and R_6, and capacitor C_3.

Capacitor C_2 couples the amplified audio signal from the audio amplifier to the three frequency selection networks. Capacitor C_4, diode D_2, and trim potentiometers V_{R_2} provide the high frequency control. Capacitors C_5 and C_6, diode D_3, and trim potentiometer V_{R_3} provide the middle frequency control. Capacitor C_7 and potentiometer V_{R_4} provide the low frequency control. The output of each frequency selection network is applied to the gate of their respective SCR. Once a certain voltage level is reached on the gate of the SCR, the SCR will turn on. The SCR will remain on until the gate voltage drops below a certain level.

General Information (continued)

The color organ control can be connected to the stereo amplifier using one of two methods. If the speaker uses an RCA phono jack, a Y-adapter will be required. If the speakers are connected to the amplifier using a twin-lead speaker wire, a wire will need to be made. The wire will require an RCA phono plug on one end and the other end will need to be stripped and tinned.

In normal use the color organ control would be connected across one of the stereo speaker inputs. A string of 120 V lamps (or any 120 V display) would be connected to each of the three jacks of the color organ control. Typically, a different color light is connected to each of the jacks. It should be noted that the total power consumption of each jack should not exceed 150 W. The master input control, potentiometer V_{R_1} is set to its minimum position. The sound is adjusted to the appropriate listening level. Potentiometer V_{R_1} is then adjusted until a pleasing display of lights occur. Trim potentiometers V_{R_2}, V_{R_3}, and V_{R_4} may need to be readjusted for the particular type of music you enjoy.

Notes

PROJECT 3

Electronic Color Organ

Name
Course
Date Due

Parts List

Resistors:

- R_1, 100 Ω 1/2 W Resistor
- R_2, 220 Ω 1/2 W Resistor
- R_3, 100 kΩ 1/2 W Resistor
- R_4, R7 1 kΩ, 1/2 W Resistor
- R_5, 10 kΩ 2 W Resistor
- R_6, 10 kΩ 1/2 W Resistor
- R_8, 470 Ω 1/2 W Resistor
- V_{R_1}, 10 kΩ 2 W Potentiometer
- V_{R_2},V_{R_3},V_{R_4}, 10 kΩ, Trim Potentiometer

Semiconductors:

- C_1,C_2, 5 μF, 50 V Electrolytic
- C_3, 100 μF, 50 V Electrolytic
- C_4, 0.005 μF, Ceramic Disk
- C_5, C_6, 0.047 μF, 100 V
- C_7, 0.47 μF, 50 V Electrolytic
- Q_1, 2N2222 or Equivalent
- D_1, D_2, D_3, 1 N4003 Diodes
- SCR_1, SCR_2, SCR_3, C106B1

Miscellaneous:

- T_1, 10 kΩ - 2 kΩ CT Audio Transformer
- S_1, SPDT Switch
- J_1, J_2, J_3, 2-prong AC Jacks
- Fuse Holder
- F_1, 5 A Fuse
- Power Cord
- RCA Phono Jack
- 12 pcs, 22 AWG Wire, 6" long
- 11,6-32x1/2" Screws
- 11, 6-32 Nuts
- 4, 6-32x1/4" Screws
- 2-1/4"x3-1/4" Printed Circuit Board
- 3-5/8"x6"x2" Box with Aluminum Cover

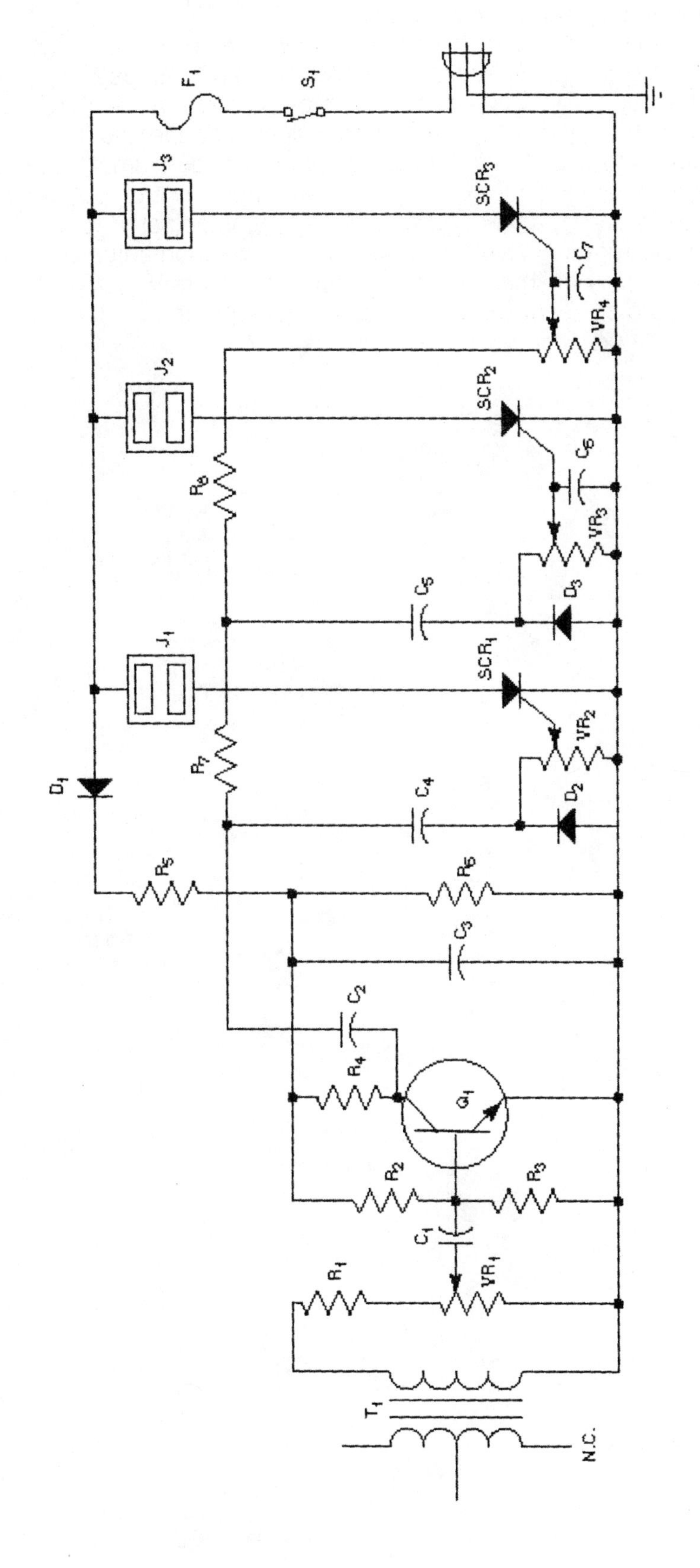

PROJECT 3

Electronic Color Organ

Notes